NCRP REPORT No. 50

Environmental Radiation Measurements

Recommendations of the
NATIONAL COUNCIL ON RADIATION PROTECTION AND MEASUREMENTS

Issued December 27, 1976

National Council on Radiation Protection and Measurements
7910 WOODMONT AVENUE / WASHINGTON, D.C. 20014

Library of Congress Catalog Card Number 76-51565
International Standard Book Number 0-913392-32-4

Preface

In recent years, the need for accurate, reliable and interpretable measurements of environmental radiation and radioactivity has increased steadily. Such requirements as the assessment of exposure to man from natural and manmade sources, the evaluation of compliance with governmental regulations, knowledge of the movement and retention of manmade radionuclides due to man's use of such materials, and the investigation of geological processes and atmospheric phenomena have all contributed to this need. This report presents a unified and systematic consideration of environmental radiation measurements, especially with respect to the identification and characterization of small radiation fields and small concentrations of specific radionuclides, and the even smaller variations or changes in them. The role of measurements in the realistic assessment of dose to man through critical pathways is emphasized, serving as an introduction to the detailed discussions on sampling and sample analysis for radioactivity. Finally, the future needs and uses of environmental measurements, of new instrumentation, and of improved interpretations of the results are presented.

The present report was prepared by the Council's Scientific Committee 35 on Environmental Radiation Measurements. Serving on the Committee during the preparation of this Report were:

James E. McLaughlin, *Chairman*

Members	*Consultants*
Bernd Kahn	Zolin G. Burson
Wayne M. Lowder	John L. Harley
Julian M. Nielsen	Carl L. Lindeken
Jacob Sedlet	Wesley L. Nicholson
McDonald E. Wrenn	Gordon K. Riel

The Council wishes to express its appreciation to the members and consultants for the time and effort devoted to the preparation of this report.

Lauriston S. Taylor
President, NCRP

Bethesda, Maryland
May 15, 1976

Contents

1. Introduction

1.1 General Considerations

Measurements of ionizing radiation and radionuclides in man's environment are required for the assessment of exposure to both natural and manmade radiation sources, determination of compliance with government regulations, and studies of the movement and retention of manmade radionuclides in environmental media and of the composition of the natural radiation environment. Measurements aid in the determination of changes in the concentrations of certain radionuclides and identification of long-term trends due to the nuclear fuel cycle, to man's use of radioactive materials, and to man's extensive modification of the earth surface. Experimental studies of natural radionuclides in the environment contribute to our knowledge of geological processes and atmospheric phenomena. All of these low-level determinations depend on reliable techniques of data gathering and analysis, supported by parallel theoretical computations.

1.2 Environmental Radiation Resulting From Man's Activities

There have been few systematic studies of environmental radiation and radioactivity, most of such efforts in the United States being largely devoted to monitoring weapons fallout and large nuclear operations. These extensive monitoring efforts shed little light on the radiation sources comprising the environment and, until recently, even on the exact manner in which local releases of radionuclides change the radiation field or nuclide concentrations in the environment. As attempts have been made to reduce the amounts of radioactive material released from nuclear facilities, it has become obvious that some previously used monitoring techniques are insufficiently sensitive and reliable to document the low levels in the environment. In this report, we identify techniques capable of either detecting relatively small changes in environmental radiation levels or uniquely identifying specific radionuclides.

For some time, the release of radioactivity from nuclear operations has been limited to assure that no member of the general population would be exposed such as to receive an annual dose of 0.5 rem above natural background and contributions from deliberate medical procedures. This assurance required measurements intended to show that resulting concentrations or doses were well below recommended values (ICRP, 1965a; 1965b). Greater emphasis is now required to show that radiation doses to man are maintained at "as low as practicable" values consistent with social and economic considerations (NCRP, 1954; FRC, 1960; NCRP, 1971). A practical consequence is that nuclear plants are being designed and built with more extensive processing of effluents. Despite these technological developments, there still will be a need for environmental monitoring programs to document radiation levels in the environs of various operations until sufficient data and experience are obtained to justify reducing or eliminating monitoring. Such documentation may also assist in improving public confidence in the operation of such facilities.

In order to meet requirements imposed by the interest in low environmental levels, measurement techniques will have to be more sensitive and more reliable than in the past. Such requirements imply greater care in designing measurement programs and incorporating more selecive, meaningful, and careful measurements rather than merely increasing the quantity and types of measurements. These measurements require greater attention to quality control and assurance.

One can identify other purposes for radiological monitoring. As man continues to modify his environment by increased reliance on the nuclear fuel cycle and by redistributing natural radioactivity, efforts should be made to monitor long-term trends in the distributions of selected radionuclides, perhaps as part of monitoring efforts for chemically toxic materials. Although many suitable measurement techniques now exist, it is not yet clear how this kind of long term, extensive monitoring can be accomplished. As a beginning, the local, regional, or global measurement of nuclides such as ^{3}H, ^{14}C, ^{85}Kr, ^{129}I, ^{137}Cs, ^{239}Pu (and related transuranic nuclides), is desirable, in ways analogous to the ^{90}Sr and ^{131}I monitoring done as a result of nuclear weapons testing. Current concentrations of these nuclides are low, but some may increase with time (UNSCEAR, 1972) and the establishment of baselines is, therefore, needed. In a similar way, the monitoring of natural radioactivity changes due to man's extensive mining, agricultural, and construction activities is desirable (Martin *et al.*, 1971; Jaworowski *et al.*, 1972). Such data can also be used to improve the assessment of small radiation doses to large segments of human populations (NCRP, 1975; NAS-NRC, 1972).

1.3 Studies In The Earth Sciences

Much of our knowledge of environmental radiation levels and the radionuclide content of environmental media has been derived from various types of geophysical or geochemical research. The specific goals of such research have determined the types of instrumentation utilized in these studies, the kinds of data generated, and the accuracy and interpretability of the results. Among such research areas have been the study of cosmic radiation in the atmosphere, the migration of radon and radon daughters in the atmosphere, and the development of radiometric techniques for uranium and thorium prospecting. The development of "practical" programs of environmental radiation and radioactivity assessment should take into account the fund of knowledge already available from research studies and the degree to which measurement programs can contribute to further research in these areas.

Possible applications of environmental gamma radiation measurements include the monitoring of snow cover, soil moisture and forest development to aid in economic planning for large regions, of atmospheric and oceanic circulation (Adams and Lowder, 1964), and of the redistribution of radioactivity. These applications require a detailed knowledge of the environmental radiation field (Kogan *et al.*, 1969). The discussion of experimental and interpretive methodologies in succeeding sections of this report will take into consideration such applications, where appropriate.

1.4 Scope

This report presents information on the properties of widely-distributed radionuclides and of typical radiation fields in the environment (Section 2), and on methods for their measurement (Sections 4–6). Emphasis is placed on the role of measurements in the realistic assessment of dose to man (see especially Section 3). Techniques applicable to routine monitoring programs during normal operation of nuclear facilities are described. Evaluations are made throughout the report of the available and developing measurement methods. Areas are identified where present knowledge is limited due to the lack of adequate measurement capabilities or systematic data collection and appraisal (Section 7). The special requirements for monitoring abnormal occurrences such as large releases of radionuclides and for occupational radiation protection are not considered.

Within the limits of this report, it is not possible to describe the details of the many measurement methods. Extensive literature citations are made so the reader may refer to methods as required. Certain relevant subjects are not treated here that are dealt with in other NCRP reports; for example, the assessment of dose to large populations from natural background radiation (NCRP, 1975) and tritium measurement methods (NCRP, 1976).

2. Natural and Manmade Environmental Radioactivity and Radiation Fields

2.1 Introduction

Environmental radiation measurement problems are commonly encountered due to complex compositions of radiation sources and fields and low radiation intensities. Interpretation of measurements is considerably aided by a prior knowledge of the characteristics of typical radiation fields and radionuclide distributions. Such knowledge includes information on the decay properties of radionuclides found in nature; the physics of radiation transport and interaction in the environment, including energy and angular distributions; and the modes of radionuclide movement.

The properties of environmental radionuclides are described here to provide a useful context for the discussions of measurement techniques and their applications in subsequent sections.

2.2 Radionuclides in Man's Environment

2.2.1 *Origin and Decay Properties*

Environmental radionuclides can be divided into three groups according to origin: (a) those of primordial origin (i.e., of sufficiently long half-life to have survived in detectable quantities since the formation of the earth), together with their radioactive daughters; (b) those continually produced by natural processes other than the decay of primordial radionuclides; and (c) those generated by man's activities. The interpretation of *in situ* and laboratory measurements of environmental radionuclides depends on knowing their half-lives, decay schemes, physical state, and distribution patterns.

The radionuclide characteristics listed in Tables 2-1 through 2-5 are the basic radioactivity decay properties of a number of naturally occurring and long-lived manmade radionuclides. With the exception

TABLE 2-1—*Properties of the thorium series*

Nuclide	Half-Life	Radiation[a]	Energy[b]	Intensity[c]	Reference[d]
			MeV	percent	
^{232}Th	1.405×10^{10} y	α_1	3.953	23	1, 2, 3
		α_2	4.012	77	
		ce_1	0.040	18	
		ce_2	0.054	5	
		e_A	0.00948	40	
^{228}Ra	5.75 y	β^-	0.055 (0.0141)	100	1
		ce	0.0054	27	
^{228}Ac	6.13 h	β_1^-	0.456 (0.132)	5.3	1, 4, 5
		β_2^-	0.494 (0.145)	6.9	
		β_3^-	0.620 (0.187)	4.6	
		β_4^-	0.996 (0.321)	7.9	
		β_5^-	1.026 (0.332)	3.5	
		β_6^-	1.125 (0.370)	11	
		β_7^-	1.180 (0.390)	39	
		β_8^-	1.752 (0.615)	14	
		β_9^-	1.819 (0.649)	2.7	
		β_{10}^-	2.089 (0.756)	6	
		ce_1	0.0368	59	
		ce_2	0.0521	21	
		ce_3	0.0743	4.6	
		ce_4	0.078	3.4	
		ce_5	0.093	1.2	
		ce_6	0.108	6.1	
		ce_7	0.123	1.9	
		ce_8	0.1635	1.4	
		e_A	0.00948	40	
		X_1	0.01297	33	
		X_2	0.09557	6.3	
		γ_1	0.1291	2	
		γ_2	0.2095	4	
		γ_3	0.2702	4	
		γ_4	0.3280	3	
		γ_5	0.3385	11	
		γ_6	0.4095	2	
		γ_7	0.4630	4	
		γ_8	0.7719	2	
		γ_9	0.7949	4	
		γ_{10}	0.833	3	
		γ_{11}	0.9111	29	
		γ_{12}	0.9645	6	
		γ_{13}	0.9689	17	
		γ_{14}	1.4592	1	
		γ_{15}	1.498	2	
		γ_{16}	1.5881	5	
		γ_{17}	1.63		
^{228}Th	1.913 y	α_1	5.341	26.7	1, 3, 4
		α_2	5.423	72.7	
		ce_1	0.0653	20	
		ce_2	0.0797	5.3	
		ce_3	0.0833	2.0	
		e_A	0.00909	21	
		X	0.01234	17	
		γ	0.0845	1.2	

TABLE 2-1—*Continued*

Nuclide	Half-Life	Radiation[a]	Energy[b]	Intensity[c]	Reference[d]
^{224}Ra	3.64 d	α_1	5.449	5.2	1, 3, 4
		α_2	5.686	94.8	
		X	0.0857	44.0	
		γ	0.24098	4.0	
^{220}Rn	55.3 s	α	6.288	100	1
^{216}Po	0.15 s	α	6.778	100	1, 3
^{212}Pb	10.64 h	β_1^-	0.154 (0.0410)	5.1	1, 4, 5, 6
		β_2^-	0.332 (0.0938)	83	
		β_3^-	0.571 (0.171)	13	
		ce_1	0.14810	30	
		ce_2	0.21009	1.2	
		ce_3	0.22223	5.3	
		ce_4	0.23462	1.3	
		e_{A1}	0.00815	20	
		e_{A2}	0.05816	1.4	
		X_1	0.01084	13.8	
		X_2	0.07892	34.0	
		γ_1	0.23862	44.9	
		γ_2	0.30009	3.4	
^{212}Bi	60.55 min	α_1	6.051	25.2	1, 3, 4, 5, 6
		α_2	6.090	9.6	
		β_1^-	0.625 (0.190)	2.2	
		β_2^-	0.733 (0.228)	1.3	
		β_3^-	1.519 (0.530)	5.0	
		β_4^-	2.246 (0.831)	54.8	
		ce_1	0.02450	19.6	
		ce_2	0.03615	4.6	
		e_A	0.00778	12.0	
		X	0.01027	7.5	
		γ_1	0.03985	1.06	
		γ_2	0.7272	7.1	
		γ_3	0.7854	1.0	
		γ_4	1.6208	1.8	
^{212}Po	0.305 μs	α	8.784	64.07	1, 3, 6
^{208}Tl	3.07 min	β_1^-	1.034 (0.341)	1.0	1, 4, 5, 7
		β_2^-	1.287 (0.440)	8.5	
		β_3^-	1.520 (0.533)	8.2	
		β_4^-	1.797 (0.647)	18.7	
		ce	0.18924	1.0	
		X	0.07674	2.3	
		γ_1	0.27735	2.4	
		γ_2	0.51080	8.5	
		γ_3	0.58314	30.5	
		γ_4	0.86037	4.6	
		γ_5	2.61466	35.93	

[a] "ce" = conversion electron; "e_A" = Auger electron.

[b] For beta particles the maximum energy is given, with the average energy in parentheses.

[c] Relative to ^{232}Th decay rate; assumed secular equilibrium.

[d] References: 1—Martin and Blichert-Toft (1970); 2—Schmorak (1970a) 3—Rytz (1973); 4—Beck (1972a); 5—Bowman and MacMurdo (1974); 6—Pancholi and Martin (1972); 7—Lewis (1971a).

TABLE 2-2—*Properties of the uranium series*

Nuclide	Half-Life	Radiation[a]	Energy[b]	Intensity[c]	Reference[d]
			MeV	percent	
^{238}U	4.49×10^9 y	α_1	4.149	23	1, 2, 3
		α_2	4.196	77	
		ce_1	0.030	23	
		ce_2	0.043		
^{234m}Th	24.10 d	β_1^-	0.100	12	1, 4, 5, 6
		β_2^-	0.101	21	
		β_3^-	0.193	67	
		ce	0.08	11	
		γ_1	0.0633	5	
		γ_2	0.0923	2	
		γ_3	0.0928	3	
^{234m}Pa	1.17 min	β_1^-	1.5	~2	1, 4, 5, 6
		β_2^-	2.29	~98	
		γ	1.0011	0.59	
^{234}U	2.48×10^5 y	α_1	4.722	27.5	1, 3, 4
		α_2	4.774	72.5	
^{230}Th	7.7×10^4 y	α_1	4.621	23.4	1, 3, 7
		α_2	4.6875	76.3	
^{226}Ra	1602 y	α_1	4.602	5.5	1, 3, 5, 6
		α_2	4.785	94.5	
		ce	0.1696	1.2	
		γ	0.1861	3.7	
^{222}Rn	3.824 d	α	5.490	100	1, 3
		X	0.08117	1.26	
^{218}Po	3.05 min	α	6.003	100	1, 3
^{214}Pb	26.8 min	β_1^-	0.20 (0.056)	2.2	1, 5, 6
		β_2^-	0.51 (0.152)	1.3	
		β_3^-	0.69 (0.215)	46	
		β_4^-	0.75 (0.236)	43	
		β_5^-	1.04 (0.348)	7	
		ce_1	0.03684	10	
		ce_2	0.04923	4.8	
		ce_3	0.15139	6.2	
		ce_4	0.20469	7.7	
		ce_5	0.22553	1.1	
		ce_6	0.26146	8.6	
		ce_7	0.27883	1.3	
		ce_8	0.33560	1.6	
		e_A	0.00815	20	
		X_1	0.01084	14	
		X_2	0.07892	22.3	
		γ_1	0.05323	2.2	
		γ_2	0.24192	7.4	
		γ_3	0.29522	17.9	
		γ_4	0.35199	35.0	
^{214}Bi	19.8 min	β_1^-	0.82 (0.264)	2.9	1, 5, 6
		β_2^-	1.08 (0.362)	6.6	
		β_3^-	1.15 (0.391)	4.9	

TABLE 2-2—*Continued*

Nuclide	Half-Life	Radiation[a]	Energy[b]	Intensity[c]	Reference[d]
(Cont'd)		β_4^-	1.27 (0.438)	4.5	
		β_5^-	1.39 (0.487)	1.6	
		β_6^-	1.43 (0.504)	8.7	
		β_7^-	1.51 (0.537)	18	
		β_8^-	1.55 (0.554)	17	
		β_9^-	1.62 (0.583)	1.2	
		β_{10}^-	1.74 (0.600)	3.5	
		β_{11}^-	1.86 (0.686)	1.0	
		β_{12}^-	1.90 (0.702)	8.6	
		β_{13}^-	3.28 (1.317)	19	
		γ_1	0.6094	43.0	
		γ_2	0.6656	1.5	
		γ_3	0.7684	4.9	
		γ_4	0.8062	1.2	
		γ_5	0.9341	3.1	
		γ_6	1.1204	14.2	
		γ_7	1.1553	1.7	
		γ_8	1.2382	6.0	
		γ_9	1.2811	1.6	
		γ_{10}	1.3778	4.6	
		γ_{11}	1.3854	1.0	
		γ_{12}	1.4017	1.6	
		γ_{13}	1.4080	2.6	
		γ_{14}	1.5095	2.2	
		γ_{15}	1.6615	1.1	
		γ_{16}	1.7299	3.0	
		γ_{17}	1.7647	15.6	
		γ_{18}	1.8477	2.2	
		γ_{19}	2.1189	1.2	
		γ_{20}	2.2045	5.0	
		γ_{21}	2.4480	1.6	
^{214}Po	162 μs	α	7.6871	100	1, 3
^{210}Pb	22.3 y	β_1^-	0.017 (0.0038)	80	1, 5, 6, 8
		β_2^-	0.061 (0.0158)	20	
		ce_1	0.03012	57.9	
		ce_2	0.04251	13.8	
		ce_3	0.04557	4.4	
		e_A	0.00815	34.5	
		X	0.01084	23.4	
		γ	0.04651	4.05	
^{210}Bi	5.012 d	β^-	1.1610 (0.3945)	100	1, 3, 8
^{210}Po	138.38 d	α	5.3045	100	

[a] "ce" = conversion electron; "e_A" = Auger electron.

[b] For beta particles the maximum is given, with the average energy in parentheses.

[c] Relative to ^{238}U or ^{226}Ra decay rates; assumed secular equilibrium.

[d] References: 1—Martin and Blichert-Toft (1970); 2—Ellis (1970a); 3—Rytz (1973); 4—Ellis (1970b); 5—Bowman and MacMurdo (1974); 6—Beck (1972a); 7—Ellis (1970c); 8—Lewis (1971b).

TABLE 2-3—*Other primordial radionuclides*

Nuclide	Abundance: Elemental	Abundance: Lithosphere[a]	Half Life	Radiation[b]	Energy[c]	Intensity[d]	Reference[e]
	percent	$\mu g\ g^{-1}$			MeV	percent	
^{40}K	0.0118	3	1.28×10^9 y	e_{A1}	0.00024	~16	1, 2
				e_{A2}	0.00296	7.0	
				β^-	1.314 (0.562)	89.3	
				X	0.00300	1.1	
				γ	1.4608	10.7	
^{87}Rb	27.85	75	4.7×10^{10} y	β^-	0.274	100	3

[a] Ref: Adams (1962)

[b] "e_A" = Auger electron.

[c] For beta particles the maximum energy is given, with the average energy in parentheses.

[d] Percentage yield relative to total radionuclide decay rate.

[e] Key to references: 1—Martin and Blichert-Toft (1970); 2—Bowman and MacMurdo (1974); 3—Verheul (1971).

of some of the cosmogenic radionuclides, the importance of which derives from their geophysical interest rather than from their abundance, only those nuclides have been included that are widely distributed in the environment with sufficient abundance to be frequent objects of environmental radiation measurement. Other nuclides may be of importance in local situations, such as ^{131}I released from nuclear facilities.

Some of the reported measurements of radiation energy and branching fraction show significant disagreements and these data have been integrated into consistent decay schemes only with some difficulty. A standard reference for the data listed is the *Table of the Isotopes*, the sixth edition of which was published in 1967 (Lederer *et al.*, 1967). Since that time, the Nuclear Data Group at the Oak Ridge National Laboratory has published updated half-life and decay-scheme data for most of the radionuclides in the tables, taking into account not only the available experimental data but also their consistency with a particular decay scheme based partly on theoretical considerations. These results have been included in the tables, except where even more recent experimental data indicate that a small modification may be in order. Thus, the energy and intensity data for gamma rays summarized by Beck (1972a) and Bowman and MacMurdo (1974) and those for alpha particles given by Rytz (1973) have been considered in choosing the values given in the tables. The differences are generally very small, one exception being the gamma energies and intensities of ^{228}Ac, which are uncertain by more than 10 percent in almost all cases.

TABLE 2-4—*Cosmogenic radionuclides*

Nuclide	Rate of Atmospheric Production of Atoms[a]	Half-Life	Radiation[b]	Energy[c]	Intensity[d]	Reference[e]
	cm^{-2} s^{-1}			MeV	percent	
^{3}H	0.25	12.35 y	β^-	0.0185 (0.00568)	100	1
^{7}Be	8.1×10^{-2}	53.3 d	e_A	0.000054	88	1, 2
			γ	0.47759	10.3	
^{10}Be	4.5×10^{-2}	2.7×10^{6} y	β^-	0.555	100	3
^{14}C	2.5	5730 y	β^-	0.1561 (0.0493)	100	1
^{22}Na	8.6×10^{-5}	2.60 y	e_A	0.00082	8.5	1
			β^+	0.5459 (0.2156)	90.49	
			γ_1	0.5110	181.08	
			γ_2	1.27454	99.95	
^{26}Al	1.4×10^{-4}	7.4×10^{5} y	β^+	1.16	84.6	2, 3
			γ_1	0.5110	169.2	
			γ_2	1.12967	2.5	
			γ_3	1.80865	99.76	
^{32}Si	1.6×10^{-4}	280 y	β^-	0.21	100	3, 4
^{32}P	8.1×10^{-4}	14.29 d	β^-	1.7089 (0.6950)	100	1
^{33}P	6.8×10^{-4}	24.8 d	β^-	0.248	100	3
^{35}S	1.4×10^{-3}	88.0 d	β^-	0.1673 (0.04879)	100	1
^{36}Cl	1.1×10^{-3}	3.01×10^{5} y	e_{A1}	0.00016	3.2	1
			e_{A2}	0.0021	1.5	
			β^-	0.7089 (0.2514)	98.1	
^{37}Ar	8.3×10^{-4}	35.0 d	γ	0.815 (max.)	100	3
^{39}Cl	1.6×10^{-3}	56.2 min	β_1^-	1.91	85	2, 3
			β_2^-	2.18	8	
			β_3^-	3.45	7	
			γ_1	0.25026	44	
			γ_2	0.98579	2	
			γ_3	1.09097	2	
			γ_4	1.2672	50	
			γ_5	1.51731	35	
^{39}Ar	4.0×10^{-3}	269 y	β^-	0.565	100	3

[a] Ref: Lal and Peters (1967).

[b] "e_A" = Auger electron.

[c] For beta particles the maximum energy is given, with the average energy in parentheses.

[d] Percentage yield relative to total radionuclide decay rate.

[e] References: 1—Martin and Blichert-Toft (1970); 2—Bowman and MacMurdo (1974); 3—Lederer *et al.* (1967); 4—Jantsch (1967).

The number of significant figures in the data taken from the indicated references does not necessarily reflect the accuracy of such data. In some cases, e.g., with many values for maximum beta particle energies, the implied accuracy refers to consistency with an assumed decay scheme rather than to absolute value.

Emissions have been included in the tables if they occur in more than 1 percent of the disintegrations. The values given for beta-

TABLE 2-5—*Widely distributed manmade radionuclides*

Nuclide	Origin[a]	Half-Life	Radiation[b]	Energy[c]	Intensity[d]	References[e]
				MeV	percent	
^{3}H	NE, NF	12.35 y	β^-	0.0185 (0.00568)	100	1
^{14}C	NE, FF	5730 y	β^-	0.1561 (0.0493)	100	1
^{54}Mn	NE	312.5 d	e_{A1}	0.00057	149	1, 2
			e_{A2}	0.00478	65.6	
			X_K	0.00547	23.6	
			γ	0.83483	99.978	
^{55}Fe	NE	2.7 y	e_{A1}	0.00063	146.6	1
			e_{A2}	0.00519	63.0	
			X_K	0.00595	25.7	
^{60}Co	NE, NF	5.26 y	β_1^-	0.31788 (0.0959)	99.92	1, 2
			γ_1	1.17321	99.92	
			γ_2	1.33248	100	
^{65}Zn	NE, NF	243.8 d	e_{A1}	0.00093	134	1, 2
			e_{A2}	0.00703	51.6	
			β^+	0.331 (0.1433)	1.41	
			X_K	0.00813	35.2	
			γ_1	0.5110	2.82	
			γ_2	1.11552	50.75	
^{85}Kr	NE, NF	10.73 y	β_1^-	0.173 (0.0475)	0.43	2, 3
			β_2^-	0.687 (0.2514)	99.57	
			γ	0.51399	0.43	
^{90}Sr-^{90}Y	NE, NF	28.5 y (Sr)	β_1^-	0.546 (0.1963)	100	1
		64.0 h (Y)	β_2^-	2.274 (0.936)	99.98	
^{95}Zr	NE	63.98 d	e_A	0.00215	1.4	3, 4
			ce	0.2164	1.1	
			β_1^-	0.3656 (0.109)	54.6	
			β_2^-	0.3981 (0.120)	44.4	
			γ_1	0.72418	44.4	
			γ_2	0.75672	54.6	
^{95}Nb	NE	35.15 d	β^-	0.1597 (0.0434)	99.92	3, 4
			γ	0.76579	99.92	
^{106}Ru-^{106}Rh	NE, NF	369 d (Ru)	β_1^-	0.0394 (0.0101)	100	1, 2
		30.4 s (Rh)	β_2^-	1.98 (0.786)	1.72	
			β_3^-	2.41 (0.986)	10.5	
			β_4^-	3.03 (1.280)	8.4	
			β_5^-	3.54 (1.525)	78.8	
			γ_1	0.5118	20.5	
			γ_2	0.6218	9.76	
			γ_3	1.0501	1.45	
^{125}Sb-^{125m}Te	NE	2.77 y (Sb)	e_{A1}	0.00319	84	1, 2
		58 d (Te)	e_{A2}	0.02272	11.1	
			ce_1	0.00365	72	
			ce_2	0.03052	9.1	
			ce_3	0.03445	1.66	
			ce_4	0.07746	12.1	
			ce_5	0.10433	9.1	
			ce_6	0.10826	2.5	
			β_1^-	0.094 (0.0246)	13.3	
			β_2^-	0.124 (0.0329)	6.0	

TABLE 2-5—*Continued*

Nuclide	Origin[a]	Half-Life	Radiation[b]	Energy[c]	Intensity[d]	References[e]
(Cont'd)			β_3^-	0.130 (0.0346)	18.8	
			β_4^-	0.241 (0.0675)	2.0	
			β_5^-	0.303 (0.0870)	39.6	
			β_6^-	0.455 (0.1381)	6.5	
			β_7^-	0.621 (0.2160)	13.6	
			X_1	0.00377	12.0	
			X_2	0.0275	73	
			γ_1	0.0355	7	
			γ_2	0.1764	7.1	
			γ_3	0.38051	1.53	
			γ_4	0.42795	30.4	
			γ_5	0.4634	10.7	
			γ_6	0.6006	18.1	
			γ_7	0.6067	5.1	
			γ_8	0.6359	11.5	
			γ_9	0.6714	1.8	
^{129}I	NF	1.57×10^7 y	β^-	0.150	100	5
			γ	0.03958	7.52	
^{134}Cs	NE, NF	2.062 y	β_1^-	0.086	27	2, 6
			β_2^-	0.210	3	
			β_3^-	0.338	7	
			β_4^-	0.410	9	
			β_5^-	0.658	61	
			β_6^-	0.69	10	
			β_7^-	0.892	1.5	
			γ_1	0.2004	4.0	
			γ_2	0.2083	4.0	
			γ_3	0.4753	1.5	
			γ_4	0.5631	8	
			γ_5	0.5692	14	
			γ_6	0.6046	98	
			γ_7	0.7958	88	
			γ_8	0.8018	9	
			γ_9	1.0384	1.1	
			γ_{10}	1.1677	1.9	
			γ_{11}	1.3650	3.4	
^{137}Cs-^{137m}Ba	NE, NF	30.0 y (Cs) 2.553 m(Ba)	e_A	0.00367	7.5	3
			ce_1	0.624204	7.80	
			ce_2	0.655656	1.42	
			β_1^-	0.514 (0.1943)	94.6	
			β_2^-	1.176 (0.4164)	5.4	
			X_1	0.03182	1.99	
			X_2	0.03219	3.67	
			X_3	0.0365	1.36	
			γ	0.661645	85.1	
^{144}Ce-^{144}Pr	NE	284.2 d(Ce) 17.28 m(Pr)	e_A	0.00408	10.3	1, 2
			ce_1	0.03813	3.3	
			ce_2	0.09154	5.3	
			β_1^-	0.1821 (0.0495)	19.4	
			β_2^-	0.2355 (0.0655)	4.2	
			β_3^-	0.3156 (0.0905)	76.3	
			β_4^-	0.810 (0.268)	1.03	
			β_5^-	2.300 (0.899)	1.22	
			β_6^-	2.996 (1.240)	97.75	

TABLE 2-5—*Continued*

Nuclide	Origin[a]	Half-Life	Radiation[b]	Energy[c]	Intensity[d]	References[e]
(Cont'd)			X_1	0.00503	2.10	
			X_2	0.03671	9.0	
			γ_1	0.08012	1.54	
			γ_2	0.13353	10.8	
			γ_3	0.69643	1.47	
^{238}Pu	SNAP, NE	87.75 y	α_1	5.4992	71.1	7, 8
			α_2	5.4565	28.7	
^{239}Pu	NE, NF	2.439×10^4 y	α_1	5.155	73.3	8, 9
			α_2	5.143	15.1	
			α_3	5.105	11.5	
^{240}Pu	NE, NF	6537 y	α_1	5.1683	76.0	8, 10
			α_2	5.1238	24.0	
^{241}Pu	NE, NF	14.8 y	β^-	0.0208	100	11
^{241}Am	NE, NF	433 y	ce_1	0.00475	8.7	1
			ce_2	0.01160	11.4	
			ce_3	0.02063	4.0	
			ce_4	0.02182	10.2	
			ce_5	0.02485	1.1	
			ce_6	0.02748	3.7	
			ce_7	0.03170	1.4	
			ce_8	0.03770	2.7	
			ce_9	0.037936	34.0	
			ce_{10}	0.053813	10.3	
			ce_{11}	0.058035	3.7	
			e_A	0.01009	36.0	
			α_1	5.3884	1.6	
			α_2	5.4430	12.8	
			α_3	5.4857	85.2	
			X	0.01394	29.0	
			γ_1	0.02635	2.5	
			γ_2	0.059536	35.9	

[a] "NE"—Nuclear explosions
"NF"—Nuclear facilities
"SNAP"—SNAP-9A (System for Nuclear Auxiliary Power) which dispersed about 1 kg ^{238}Pu in the earth's atmosphere (Hardy *et al.*, 1973)
"FF"—Fossil fuel power plants and other industries.

[b] "ce" = Conversion electron; "e_A" = Auger electron; "X_K" = K x ray.

[c] For beta particles the maximum energy is given, with the average energy in parentheses.

[d] Percentage yield relative to total radionuclide decay rate.

[e] References: 1—Martin and Blichert-Toft (1970); 2—Bowman and MacMurdo (1974); 3—Martin (1973); 4—Medsker and Horen (1972); 5—Horen (1972); 6—Nuclear Data Group (1965); 7—Ellis (1970a); 8—Rytz (1973); 9—Artna-Cohen (1971); 10—Schmorak (1971); 11—Ellis (1970b).

particle energy are maximum energies; the values in parentheses are average energies given by Martin and Blichert-Toft (1970).

Tables 2-1 and 2-2 give the properties of radionuclides comprising the thorium and uranium series, which, along with ^{40}K, contribute most to natural radioactivity and the consequent environmental radiation field. The actinium series, which begins with ^{235}U (about 0.72 percent as abundant as ^{238}U and with only 4.6 percent of the activity), is a negligible contributor to natural background. The neptunium series begins with ^{237}Np, the half-life of which is too short relative to the earth's age to exist as a primordial nuclide, although trace quantities are generated by cosmic-ray interactions with ^{238}U. The actinium and neptunium series are not considered further in this report.

The intensities in the tables are for assumed secular equilibrium among the member nuclides, but the arrangement according to nuclide permits one to correct for non-equilibrium situations, if the degree of disequilibrium is known.

Most of the gamma rays from the uranium series are emitted during the decay of ^{214}Pb and ^{214}Bi, and the principal contributor to external dose rate is the latter. Similarly, ^{208}Tl in the thorium series contributes most to the dose rate. There is disagreement, particularly in the older literature, on the intensities of some of the more energetic emissions from these and other nuclides in the two series. Because of the importance of those data in environmental radiation studies, more precise intensity values are desirable.

Other primordial radionuclides are listed in Table 2-3, including ^{40}K, which is a main contributor to naturally occurring radioactivity and to the observed background gamma radiation field.

The naturally occurring cosmogenic radionuclides that are continuously generated by cosmic-ray interactions in the atmosphere are listed in Table 2-4. Minute quantities of other radionuclides are also produced in the ground by neutron interactions and by the spontaneous fission of the heavy natural radionuclides, particularly ^{238}U. The rate of such production per unit surface area of ground is less than 10^{-6} atoms cm^{-2} s^{-1}.

Detectable quantities of manmade radionuclides are widely distributed, particularly as a result of nuclear weapons testing in the atmosphere. These nuclides and others that may be released from nuclear power fuel cycles are listed in Table 2-5. Some of these, such as ^{14}C and ^{3}H, are also produced by natural processes. In such cases, care must be exercised in determining the manmade contributions and their origin.

Uncertainties also exist in the decay schemes of manmade nuclides. Thus, although the values in Table 2-5 can be considered as the best available from published data, revisions will doubtless be made, as in the case of the natural emitters in previous tables.

Other manmade radionuclides not listed in Table 2-5 are found near nuclear facilities or weapons testing areas, or are widely distributed but have relatively short half-lives. Among the former are ^{41}Ar, ^{138}Xe, ^{133}Xe, ^{87}Kr, ^{88}Kr, ^{138}Cs, ^{88}Rb, and ^{131}I that are released by some nuclear facilities in measurable quantities and contribute most of the additional nearby dose rate (e.g., Beck *et al.*, 1972a). Among the latter are ^{140}Ba–^{140}La, ^{124}Sb, and ^{103}Ru, as well as ^{41}Ar and ^{131}I.

2.2.2 *General Distribution Patterns*

A common feature of many environmental radiation measurement programs is the study of radionuclide distributions and concentrations. Information of this type has been accumulated by many investigations directed toward a variety of goals, but only in recent years has it been put together in a coherent fashion because of the practical need for a quantitative assessment of man's perturbations of the radiation environment. The considerable but scattered literature has been summarized by the United Nations Scientific Committee on the Effects of Atomic Radiation in a number of reports, most recently in 1972 (UNSCEAR, 1972). Earlier references include Lowder and Solon (1956) and Klement (1965).

2.2.2.1 *Lithosphere*. Potassium-40 and the radionuclides of the uranium and thorium series contribute most of the naturally-occurring radioactivity in rocks. Potassium-40 constitutes 0.0118 percent of natural potassium, which in turn constitutes about 2.6 percent of the accessible lithosphere (Adams, 1962). The resulting average abundance by weight of ^{40}K is comparable to that of uranium and about one-fourth that of thorium (Adams *et al.*, 1959). Rubidium-87 is considerably more abundant than any of these nuclides, but is a much less significant nuclide because of its long half-life and low-energy beta-particle emission.

Although the radionuclide content of rocks is a complicated function of their geochemical history, varying considerably among the various types, certain generalizations can be made that derive from extensive geological investigations. For example, the radioactivity in igneous rocks is related to the quantity of silicates, being highest in acidic varieties and lowest in the ultrabasic rocks (e.g., dunites). Igneous rocks generally exhibit higher radioactivity than sedimentary rocks, while metamorphic rocks have concentrations typical of the unmetamorphosed rocks from which they are derived. Certain sedimentary rocks, including some shales and phosphate rocks, are highly radioactive, while other types, notably limestone and various evaporites (e.g., halite, anhydrite, and gypsum), are quite low in radionuclide content. Table 2-6 shows typical natural radioactivity concentrations in common rocks. These values and those for other media in the following tables should be regarded as approximate expectation values.

The radioactivity of soil, usually a more direct determinant of radiation levels in the outdoor environment, depends not only on that of the parent rock (which may not be identical with the local bedrock) but also on soil formation and transport processes. Typical uranium, thorium, and total potassium contents of a wide variety of soils in North America and Europe are 2 $\mu g\ g^{-1}$, 8 $\mu g\ g^{-1}$ and 1.5 percent, respectively, though observed contents are a strong function of soil type and soil horizon (Baranov and Morozova, 1973). Thus, significant variations of soil radioactivity with location and depth are common. Table 2-7 lists typical *in situ* soil concentrations of the natural radionuclides.

The relatively few simultaneous measurements of the radium and uranium contents of soil indicate that radioactive equilibrium is roughly attained in many soils, but large deviations from equilibrium are also observed due to different geochemical properties of radium and uranium compounds. Departure from equilibrium occurs even

TABLE 2-6—*Radionuclides in rocks*

Type of Rock	K	U	Th
	percent	ppm	ppm
Igneous			
Silica (e.g., granites)	3.3	4.7	20.0
Intermediate (e.g., diorites)	2.3	1.8	8.0
Mafic (e.g., basalt)	0.8	0.9	2.7
Ultramafic (e.g., dunites)	0.5	0.03	6.0
Sedimentary			
Limestones	0.3	2.2	1.7
Carbonates	—	2.1	1.9
Sandstones	1.2	~1.5	~3.0
Shales	2.3	3.5	11.0
Mean value (Earth's crust)	2.3	3.0	11.4

References: Adams (1962); Vinogradov (1959).

TABLE 2-7—*Radionuclides in soil*

Radionuclide	Soil concentration: Typical range	Soil concentration: World average	Mean specific activity
	$g\ g^{-1}$ soil	$g\ g^{-1}$ soil	$Ci\ g^{-1}$
^{40}K	$(0.5–3.0) \times 10^{-6}$	1.5×10^{-6}	1.0×10^{-11}
^{87}Rb	—	4.0×10^{-5}	3.5×10^{-12}
^{226}Ra	$(0.5–2.0) \times 10^{-12}$	8.0×10^{-13}	8.0×10^{-13}
^{232}Th	$(2–12) \times 10^{-6}$	6.0×10^{-6}	6.5×10^{-13}
^{238}U	$(1–4) \times 10^{-6}$	2.0×10^{-6}	6.7×10^{-13}

References: Vinogradov (1959); Grodzinskii (1965); Baranov and Morozova (1973).

more readily for those ^{238}U daughters beyond ^{222}Rn because of the escape of gaseous radon from the soil matrix into the pore spaces and subsequent migration elsewhere prior to decay. This phenomenon is much less marked in the ^{232}Th series because of the shorter half-life of gaseous ^{220}Rn.

The mean soil content of ^{87}Rb is 40 μg g^{-1}, which results in a beta activity of the order of 10 percent that of ^{40}K.

Two manmade radionuclides widely distributed in near-surface soils are ^{90}Sr and ^{137}Cs deposited by fallout from nuclear weapons tests. Though their geographic and depth distribution patterns are somewhat irregular, most of each nuclide is generally retained in the upper 15 cm of soil, with the concentrations usually decreasing roughly exponentially with depth. The ^{90}Sr and ^{137}Cs concentrations near the soil surface are strongly time dependent, because of their variable deposition rates over many years, and their gradual depletion by decay, erosion and leaching.

Given the typical soil contents of the natural radionuclides indicated above, it can be inferred that the natural alpha-particle activity of soils is contributed by the thorium and uranium series in about a 2:1 ratio. Potassium-40 accounts for at least one-half of the natural beta activity, with the two series plus ^{87}Rb making roughly comparable contributions to the remainder. Potassium and the thorium series each contribute about 40 percent of the natural gamma-emission rate from soil, with the uranium series accounting for the remaining 20 percent. The manmade nuclides ^{90}Sr and ^{137}Cs are present in sufficient quantities to contribute significantly to the total soil activity. Their beta-activity concentrations in surface soil have recently been comparable to that of ^{87}Rb ($\sim$1 pCi g^{-1}), and the gamma-activity concentration of ^{137}Cs has been approximately one-half that of the uranium series. Thus, the contribution of fallout radionuclides to the total beta or gamma activity of surface soils is now about 10 percent.

The environmental radiation inside buildings from sources in the lithosphere can differ significantly from that in the nearby out-of-doors environment. In general, two competing effects are observed. The building provides shielding against outdoor environmental radiation, but the building material itself is an additional radiation source. Oakley (1972) has summarized the few studies of indoor radiation in the literature, and Hultqvist (1956) and Hamilton (1971) have reported radioactivity concentrations in building materials in Sweden and the United Kingdom, respectively. On the average, indoor gamma-radiation levels are comparable to those in the outdoor environment.

2.2.2.2 *Atmosphere.* The radionuclides normally found in ground-level air include gaseous radon from the uranium and thorium series, along with their decay products, cosmogenic radionuclides produced in the atmosphere, and various fission and activation products. The most significant of these usually are ^{222}Rn and its daughters.

Several tens of percent of the radon atoms escape from a typical soil matrix into the soil air and diffuse from the upper layers into the atmosphere in a manner that depends critically on local meteorological conditions. Jacobi (1963) showed that, within a few meters of the ground, the ^{220}Rn radioactivity concentration is not much less than that of ^{222}Rn, but it declines much more rapidly with height, under typical atmospheric conditions. The concentration of ^{212}Pb, the relatively long-lived daughter of ^{220}Rn, is much lower at ground level than that of ^{222}Rn, but its concentration is nearly constant to a height of several hundred meters. These phenomena reflect the differing time scales of the production and decay of the various nuclides and of atmospheric transport processes, and have been confirmed by many measurements.

From the standpoint of measurement, the concentration of the radon daughter nuclides is more significant than that of the radon itself. The relatively short-lived daughters of ^{222}Rn, i.e., ^{214}Pb and ^{214}Bi, contribute most of the observed beta- and gamma-radiation flux densities from atmospheric radionuclides. By producing a washout of the aerosols to which the daughter nuclides are usually attached, precipitation can cause a large short-term reduction in the atmospheric radon daughter content. The differences between aerosol and gaseous radon movement frequently produce a disequilibrium between ^{222}Rn and its short-lived daughters in the near-ground air, with the daughter concentrations being somewhat lower. This phenomenon should be taken into account when inferring radon concentrations from daughter measurements.

The radon content of ground-level air varies considerably with location and time. The half-life of ^{222}Rn is sufficiently long so that it can be transported far from its place of origin prior to decay. As a result, the observed radon concentrations at any location may not be closely related to the degree of exhalation from the ground nearby.

Both short-term periodic measurements and long-term measurements are affected by time variations of ^{222}Rn and its daughters in the near-ground air. Though these variations are complex, the effect of wind and atmospheric stability, and the water content of the soil, on the transport of radon have been well documented. Significant diurnal and seasonal variations of ^{222}Rn concentrations are observed and can be mostly explained by changes in atmospheric stability conditions. The diurnal ^{222}Rn cycle seems to be affected both by atmospheric and soil conditions, though both radon nuclides tend to exhibit maximum ground-level concentrations in the early morning hours when the most stable conditions tend to occur (Israelsson *et al.*, 1972). Gold *et al.* (1964) and Cox *et al.* (1970) reported higher mean ^{222}Rn levels at Cincinnati in the fall months as illustrated in Figure 2-1, produced by seasonal variations in average stability conditions. The usual annual cycle of mean ^{220}Rn ground-level air concentrations is different from that for ^{222}Rn, showing a maximum in the summer. This is attributed to the short half-life of ^{220}Rn and the resulting short distance that it migrates prior to decay, and thus a greater sensitivity to local soil porosity and water content, which influence the emanation rate, than to the large-scale atmospheric conditions which affect ^{222}Rn transport.

Because of their relatively long half-lives, the ^{210}Pb, ^{210}Bi and ^{210}Po daughters of ^{222}Rn have quite different distribution patterns in the atmosphere than radon and its shorter-lived daughters. Ground-level air concentrations are usually several orders of magnitude below the equilibrium values, because of the combined effect of vertical diffusion to higher altitudes and deposition on the ground by aerosol

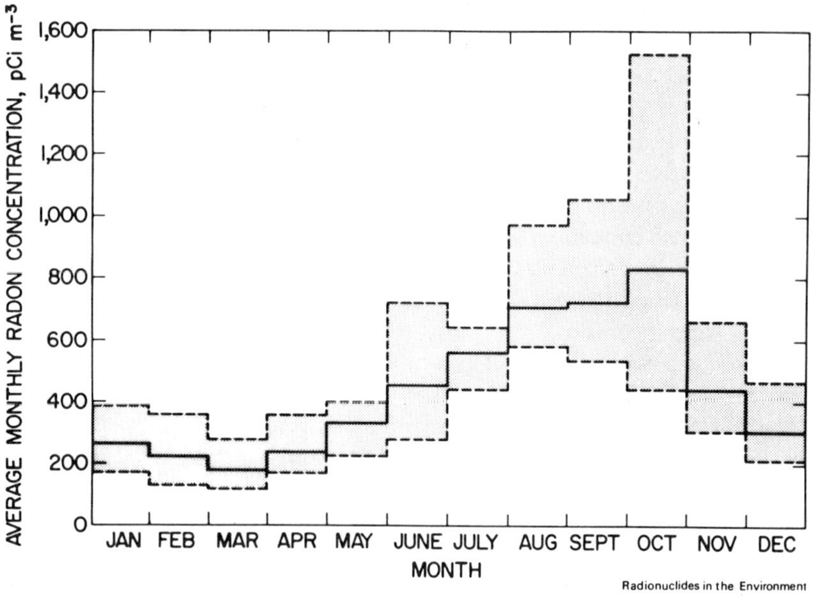

Fig. 2-1. Mean and extreme average monthly morning ^{222}Rn concentrations, 1959–1966, Cincinnati, Ohio, determined from alpha counting of daughter products collected on filters [from Cox, W. M., Blanchard, R. L. and Kahn, B. (1970). "Relation of radon concentration in the atmosphere to total moisture detention in soil and atmospheric thermal stability," page 436 in *Advances in Chemistry Series No. 93* (American Chemical Society, Washington), by permission].

settling or precipitation. Typical surface air concentrations of these natural nuclides, and the potentially important ^{85}Kr and the interesting cosmogenic ^{7}Be, are shown in Table 2-8.

The air concentrations of fallout radionuclides are a strong function of the history of atmospheric nuclear explosions. Nowadays the observed fallout in the troposphere derives from the long-term stratospheric inventory and a few recent atmospheric tests. As in the case of the cosmogenic radionuclides produced primarily in the stratosphere, fallout exhibits an annual variation in ground-level concentration, with a maximum in the late spring in the northern hemisphere.

The air concentration of radon in enclosed, indoor environments depends on the radium content and porosity of the building material and the degree of ventilation of the indoor air. Daily and seasonal patterns would therefore be expected that are closely coupled to human living habits and poorly related to those observed outdoors. Under conditions of poor ventilation, the radon levels can be several orders of magnitude higher than the typical values in Table 2-8.

2.2.2.3 *Hydrosphere*. The concentrations on naturally-occurring radionuclides in water are several orders of magnitude less than those in rocks and soils and are mostly due to ^{40}K. In many natural waters, there is a significant shift away from equilibrium between parent and daughter nuclides in the uranium and thorium series. Elemental

TABLE 2-8—*Radionuclides in the atmosphere*

Radionuclide	Surface air content	
	Typical range	Mean value
	pCi m^{-3}	pCi m^{-3}
Uranium series:		
^{222}Rn	20–500	120
^{214}Pb	0–500	100
^{214}Bi	0–500	100
^{210}Pb	0.003–0.03	0.01
^{210}Po	–	0.003
Thorium series:		
^{220}Rn	–	100
^{212}Pb	0.5–10	2
Others:		
^{85}Kr	–	17[a]
^{7}Be	0.02–0.20	0.06

[a] 1972 value; gradually increasing from ~2 pCi m^{-3} in 1960 to an anticipated value of ~100 pCi m^{-3} in the 1980's (Kirk, 1972).

References: Jacobi (1963); Peirson (1963); Lockhart (1964); Patterson and Lockhart (1964); Peirson *et al.* (1966); Malakhov and Chernysheva (1965); Vilenskii *et al.* (1965); UNSCEAR (1972); Jaquish and Johns (1972).

uranium and radon daughters are frequently observed in excess concentrations in the water relative to radium, while radium and thorium are strongly concentrated in the bottom sediments. The radionuclide content of sea water exhibits a fairly narrow range, unlike those for bodies of fresh water and urban water supplies, which are more dependent on local conditions of rock and soil radionuclide content and geochemistry.

The content of fallout radionuclides in sea water varies with geographical location and time in a complex manner. The existing data have been summarized in a National Academy of Sciences report (NAS-NRC, 1971).

2.2.2.4 *Biosphere*. The most abundant radionuclide in the biosphere is ^{40}K, the average potassium content of plants being roughly 0.05 percent and of animal tissues 0.2 percent. The contents of the various radionuclides in the uranium and thorium series are highly variable and they almost always are not in equilibrium. Radium is preferentially taken up by plants relative to uranium, while the gaseous radon daughters of radium escape almost completely. These characteristics indicate the importance of knowing the mechanism of nuclide transfer in plants and animals for dose calculation. Table 2-9 presents typical natural radionuclide concentrations in the biosphere which, as in previous tables, are merely indicative.

The fallout nuclides, ^{137}Cs and ^{90}Sr, enter plants by direct deposition as well as uptake from the soil, but, except during periods of high deposition rates such as occurred in the early 1960s, their contributions to total plant radioactivity are small. One useful summary of the literature through 1968 was given by Pertsov (1973). The acquisition of more extensive data on radionuclide concentrations throughout the biosphere is probably needed if the large numbers of pathways to man are to be examined quantitatively.

TABLE 2-9a—*Natural radionuclide content of plants and animals*[a]

Sample type	Concentration			
	K	U	Ra	Rb
	percent	μg g⁻¹	μg g⁻¹	μg g⁻¹
Grain cultures (dry)	0.9	—	—	—
Fodder grasses (dry)	1.3	—	—	—
(wet)	0.35	—	—	—
Mixed forest (dry)	0.11	—	—	—
(wet)	0.05	—	—	—
Coniferous forest (dry)	0.12	—	—	—
(wet)	0.06	—	—	—
Plants (general)	0.05	10^{-4}–10^{-1}	10^{-2}	1–10
Animals	0.2	10^{-2}–10^{-1}	10^{-3}	20–40

TABLE 2-9b—*Total natural radioactivity in plants*[a]

Radiation	Concentration
	pCi g^{-1}, gross weight
α	0.14–3.1[b]
β	7.8–123[c]
^{40}K	1–50
^{87}Rb	~0.1
^{210}Po	0.01–10

[a] References: Belousova and Shtukkenberg (1961); Grodzinskii (1965); Klement (1965); Kogan *et al.* (1969).

[b] Mainly as ^{210}Po; other U + Th series nuclides.

[c] Mainly as ^{40}K; ^{210}Pb; ^{210}Bi; other U + Th series nuclides.

2.3 Environmental Radiation Fields

2.3.1 *General Properties*

In situ measurements are difficult because of the complex composition of the radiation field and the low intensities of the various components. The complexity of the field is a consequence of the many natural and manmade sources, as well as the presence of the highly energetic charged particles from cosmic ray interactions in the atmosphere. Table 2-10 gives the characteristics of a typical radiation field

TABLE 2-10—*A typical environmental radiation field (one meter height)*

Radiation	Energy	Source	Absorbed dose rate: Free air	Absorbed dose rate: Gonads[a]
	(MeV)		μrad h^{-1}	μrad h^{-1}
α	1–9	radon (atm)	2.7	~0
β	0.1–2	radon (atm)	0.2	~0
	0.1–2	K, U, Th, Sr (soil)	2.5	~0
	2–200	cosmic rays	0.7	0.3
γ	≤2.4	radon (atm)	0.2	0.1
	≤1.5	K (soil)	2.0	1.3
	≤2.4	U (soil)	1.0	0.7
	≤2.6	Th (soil)	2.4	1.8
	≤0.8	Cs + other fallout (soil)	0.3	0.2
n	0.1–100	cosmic rays	0.1	0.2
p	10–2000	cosmic rays	0.1	0.1
μ	100–30,000	cosmic rays	2.3	2.3
Total:			14.5	7.0

[a] Reference: UNSCEAR (1972).

at sea level in the continental United States. The absorbed dose rates in free air are about 6 μrad h^{-1} from terrestrial gamma rays and 3.2 μrad h^{-1} at sea level from cosmic rays (Beck *et al.*, 1966; Oakley, 1972). Estimates of the respective absorbed dose rate contributions to the reproductive organs of an individual standing at such a location are given in Table 2-10 (UNSCEAR, 1972). The comparison of the relative contributions to dose rate illustrates the dependence of radiation response on the properties of the "detector" and its surroundings. A measurement of total absorbed dose rate in free air, or even that from only the penetrating components (cosmic and gamma rays), such as is measured with an ionization chamber, does not yield a quantity that is simply related to gonad absorbed dose rate.

2.3.2 *Cosmic Radiation*

2.3.2.1 *Composition*. The cosmic-ray charged particles in the atmosphere are almost entirely secondaries produced by nuclear interaction higher in the atmosphere. The primary galactic radiation incident on the atmosphere has the composition given in Table 2-11. These particles undergo nuclear interaction with air nuclei in the upper atmosphere and produce secondary protons (p) and neutrons (n), as well as charged and uncharged pions (π). The secondary nucleons in turn generate additional pions by the nucleonic cascade interactions. The short-lived pions are the immediate progenitors of the particles that predominate in the lower atmosphere, the charged

TABLE 2-11—*Composition of primary galactic cosmic rays at high latitudes*[a]

Atomic number	Vertical particle flux ≥1.3 GV[b]	
	Solar minimum	Solar maximum
	m^{-2} sr^{-1} s^{-1}	
1	2000	800
2	300	120
3–5	6	2
6–9	16	6
10–20	6	2
20 or more	2	1

[a] In the continental United States.

[b] Volt is a unit of magnetic rigidity, a quantity related to the deflection of a charged particle from a straight path by the earth's magnetic field; such deflection depends on particle momentum. 1.3 GV corresponds to kinetic energies of 0.66 GeV for protons and close to 1.3 GeV for the heavier particles.

Reference: Webber (1967).

pions decaying into muons (μ) and the uncharged pions decaying into a pair of photons (γ).

Many of these photons are sufficiently energetic to produce high-energy electrons (e), which in turn generate more photons by annihilation and bremsstrahlung production, the resulting multiplication of the electron-photon flux densities being called the electromagnetic cascade. A similar cascade can be generated by muon decay into electrons.

The principal cosmic-ray nuclear reactions can be summarized as follows (with υ representing neutrinos):

(a)	$p + \text{air} \rightarrow p + n + \pi^{\pm} + \pi^{\circ}$	nucleonic cascade
(b)	$n + \text{air} \rightarrow p + n + \pi^{\pm} + \upsilon^{\circ}$	nucleonic cascade
(c)	$\pi^{\pm} \rightarrow \mu^{\pm} + \gamma$	muon production via pion decay
(d)	$\pi^{\circ} \rightarrow 2\gamma \rightarrow 4e^{\pm} \rightarrow \ldots$ etc.	pion decay to electromagnetic cascade
(e)	$\mu^{\pm} \rightarrow e^{\pm} + 2\upsilon\ (\rightarrow \gamma \rightarrow \ldots$ etc.)	muon decay to electromagnetic cascade

The most important components of this complex of reactions in the lower atmosphere are the muons from reactions (c) and the electrons from reactions (e) and muon ionization of the air atoms. At higher altitudes (>3000 km), the electrons from reactions (d) are the dominant charged particle. Protons and pions from reactions (a) and (b) form only a small proportion of the charged particle flux at atmospheric depths greater than 100 g cm^{-2}. Neutrons from (a) and (b) are not important in terms of absorbed dose rate in free air except at high altitudes, although they become important contributors to the dose equivalent rate. Typical amplitudes of the various types of cosmic-ray variations are indicated in Section 2.3.2.3 in the context of available experimental data.

2.3.2.2 *Space and Time Variations*. The composition and intensity of the cosmic radiation field at a particular location and time depend on a number of factors, the most significant of which are the following:

(a) the approximately constant galactic cosmic-ray particle flux,
(b) the modification of the galactic particle flux by the time- and space-varying magnetic fields trapped in the solar wind ("modulation"),
(c) the modification of the primary cosmic-ray spectrum in the vicinity of the earth by the geomagnetic field ("latitude effect"),
(d) the enhancement of particle fluxes near the earth produced by radiation emitted from the sun during solar flares,

(e) the mass thickness of air above the location ("altitude or pressure effect"), and

(f) the spatial distribution of air mass above the location ("atmospheric temperature effect").

Factors (b)–(d) are extra-atmospheric effects that influence the "source" intensity and energy spectrum of primary particles of galactic and solar origin incident on the atmosphere, while factors (e) and (f) are effects of the atmosphere on the propagation of cosmic-ray secondaries to the spatially-fixed measurement point.

The solar modulation of the galactic cosmic-ray particles entering the solar system is presently an active field of investigation. The literature up to 1971 has been reviewed by Jokipii (1971) and Rao (1972). Modulation is produced by the scattering of the galactic primary particles on the disordered magnetic fields trapped in the solar plasma. This plasma consists of low energy solar electrons and protons moving radially outward from the sun to great distances and is closely related to the excitation of the solar corona. While time variations in the galactic primaries incident on the earth's atmosphere are inversely correlated with the 11-year solar activity cycle as measured by the sunspot number (Forbush, 1958; Neher and Anderson, 1965) and by the coronal excitation (Pathak and Sarabhai, 1970), the actual mechanisms producing the modulation effect are not yet well understood (Mathews *et al.*, 1971).

Short-term solar flares can produce large changes in the particle flux density incident on the earth's atmosphere. Such flare activity produces two partly competing effects, an initial enhancement followed by a general depression of the total flux density.

The enhancement, frequently observed approximately 15 minutes after a major solar flare event, is mostly due to protons with energies greater than 100 MeV along with some electrons and heavier nuclei that are emitted by the flare. The total particle flux densities of energies greater than 20 MeV from such events can be orders of magnitude greater than those of galactic cosmic rays. However, the energy spectrum of solar particles is more degraded (NCRP, 1975) and the galactic particles always dominate at energies greater than a few GeV. Although the large short-term increases are of some concern in terms of the exposure of passengers in high-altitude aircraft (ICRP, 1966), their effect decreases with depth much more rapidly than the galactic secondary cosmic ray intensity. Normally, the enhancement of the muon component and hence the total ionizing component in the latitudes of the continental United States ranges from undetectable up to about 10 percent during solar flare activity. For an unusually large and energetic event such as that of February 23, 1956, the increase of the muon intensity can be up to several times normal for a period of a few hours (Dorman *et al.*, 1956).

The second effect, the Forbush decrease in galactic cosmic-ray intensity, occurs 1 to 2 days after a solar flare and may amount to a decrease of 10 percent. The total period of increased modulation generally is a few days, although full recovery to pre-flare modulation levels may take several weeks. In a few cases, the post-decrease modulation level has been different from the pre-decrease level. Lockwood (1971) and Lockwood *et al.* (1972) have suggested that the Forbush decreases can be regarded as an integral part of the 11-year modulation variations. Such decreases are frequently noted without any preliminary enhancement of cosmic-ray intensity, an indication that only low-energy solar particles are involved.

The earth's magnetic field exercises an additional modulating effect on interplanetary primary particles by preventing particles below particular energies from reaching a given point at the top of the atmosphere at given angles of incidence. These cut-off energies range from 0 at the magnetic poles to about 16 GeV at the geomagnetic equator for protons incident in the vertical direction. Tables of these cut-off energies in the vertical direction have been published by Shea and Smart (1967) and Shea *et al.* (1968).

The atmospheric pressure effect is a shielding phenomenon. The initial buildup in particle fluxes at high altitudes due to the production of a multiplicity of secondary particles per unit primary particle in both the nucleonic and electromagnetic cascades is followed by a gradual attenuation as the energies of the secondary particles decline below those at which multiple secondary particle production takes place. The day-to-day variations of up to 5 percent in the barometric pressure produce somewhat greater changes (<10 percent) in cosmic-ray intensity in the lower atmosphere.

It is important to note that the "altitude profile" of cosmic-ray intensity is actually a pressure- or atmospheric-depth (g cm^{-2}) profile, and the conversion of pressure data to altitudes assumes a "standard" atmosphere that only approximately resembles the real atmosphere at any particular time. Therefore, cosmic-ray intensity data are frequently given as functions of atmospheric pressure or depth, and the value appropriate to any field measurement can be most accurately inferred from a determination of the barometric pressure at that time.

The atmospheric temperature effect primarily derives from the influence of the linear dimensions of the atmosphere on the probability of muon decay prior to its reaching a fixed spatial location. Muon production is a function of atmospheric depth, while the muon decay probability, at a particular initial velocity (or energy), is a function of time (approximately proportional to linear path length) rather than mass thickness traversed. An increase in atmospheric temperature causes expansion of the atmosphere that in effect displaces the muon source further from a detector and reduces the flux of muons and their secondaries seen by the detector. The temperature effect is decoupled from the pressure effect in terms of its causes. Both diurnal and seasonal variations in cosmic-ray intensity have been observed that can be attributed to atmospheric temperature changes. The diurnal variations in muon intensity, of the order of 0.1 percent in amplitude, are negligible as compared with the 5 percent seasonal effect.

The atmospheric pressure and temperature effects have been discussed in detail by Wada (1960); Carmichael *et al.* (1967); and Dorman (1972), among others. These effects are of particular interest in studies of primary cosmic ray properties with groundbased monitors.

2.3.2.3. *Intensity.* Many measurements of cosmic-ray intensity (e.g., ionization rate, absorbed dose rate, particle flux density) in the atmosphere have been made. However, these are difficult to compare exactly because of the complicated dynamics of both the cosmic-ray field within the atmosphere and the incident primary particles. For most environmental radiation measurements, it is sufficient to have available estimates within an accuracy of ±5–10 percent of the mean composition, particle flux densities, air ionization rates, and tissue dose rates for the cosmic-ray field as a function of height in the atmosphere.

Table 2-12 gives the properties of the cosmic-ray field at sea level at 50°N geomagnetic latitude (Lowder and O'Brien, 1972) derived from both experiments and calculations (O'Brien and McLaughlin, 1972; O'Brien, 1972). The indicated ranges show variations over the 11-year solar activity cycle.

TABLE 2-12—*Sea level cosmic-ray field*

Total charged particle flux density	$2.4\text{–}2.6 \times 10^{-2}$ cm^{-2} s^{-1}
Muon flux density	$\sim 1.8 \times 10^{-2}$ cm^{-2} s^{-1}
Electron flux density	$\sim 0.7 \times 10^{-2}$ cm^{-2} s^{-1}
Neutron flux density	$6.5\text{–}6.9 \times 10^{-3}$ cm^{-2} s^{-1}
Total charged particle ionization rate, I[a]	2.0–2.2 cm^{-3} s^{-1} (STP)
Total absorbed dose rate in free air	26–30 mrad y^{-1}
Maximum neutron dose-equivalent rate[b]	6.5 mrem y^{-1}

[a] See definition in text.

[b] At 5-cm depth in 30-cm thick semi-infinite tissue-equivalent slab irradiated isotropically from both sides.

Reference: Lowder and O'Brien (1972).

At sea level, nearly 75 percent of the ionization rate derives from muon collision electrons, 15 percent from muon decay electrons, and 10 percent from electrons, protons and neutrons produced in the nucleonic cascade. This last component increases with altitude more rapidly than the muon component and becomes comparable to the latter at a depth of 700 g cm^{-2} (O'Brien, 1972).

A frequently-used measure of the intensity of the total charged particle component of the cosmic-ray field is the free-air ionization rate, I (ion pairs cm^{-3} s^{-1} at STP). Many ionization chamber measurements of this quantity as a function of altitude and latitude have been made, particularly by Neher (1971); Nerurkar and Webber (1964); and Lowder *et al.* (1971). Small discrepancies among the various measurements have been discussed by Lowder and Beck (1966); George (1970); Raft *et al.* (1970); Carmichael (1971); and Liboff (1972).

Figure 2-2 is based on ionization-rate data from Lowder and Beck (1966) for the lower atmosphere, which agree closely with the measurements of Shamos and Liboff (1966) and Liboff (1972). These data can be considered as average values over the 11-year solar cycle for 50°N geomagnetic latitude. Note that the slope of this "altitude profile" closely approximates the variation in free-air ionization rate per unit pressure change at a measurement location on the earth's surface. The conversion of pressure to altitude in Figure 2-2 was accomplished by use of the tables for the U.S. Standard Atmosphere (1962).

Experiments with shielded and unshielded ionization chambers at ground and balloon altitudes, and with muon and neutron monitors, provide information on the nature of the variations in muon, total charged-particle, and neutron intensity. These variations may not be negligible in terms of measurements of environmental radiation, as indicated in Table 2–13.

The geomagnetic latitude variation of the sea-level muon (and thus total charged particle) ionization rate due to the modulating effect on the primary radiation of the earth's magnetic field is illustrated in Figure 2–3. The cosmic-ray absorbed dose rate gradually declines to 90 percent of its high latitude value between 40° and the geomagnetic equator (Carmichael and Berkovitch, 1969).

The cosmic-ray intensity data given above refer specifically to the outdoor environment. Indoors, the cosmic-ray field is somewhat reduced, the extent of which depends on the composition and thickness of structural shielding. However, the muons that dominate at ground altitudes are highly-penetrating particles, as can be seen from the cosmic-ray intensity profile in Figure 2–2 (~10 percent attenuation in 50 g cm^{-2} of air at sea level). Muon attenuation is less in dense than

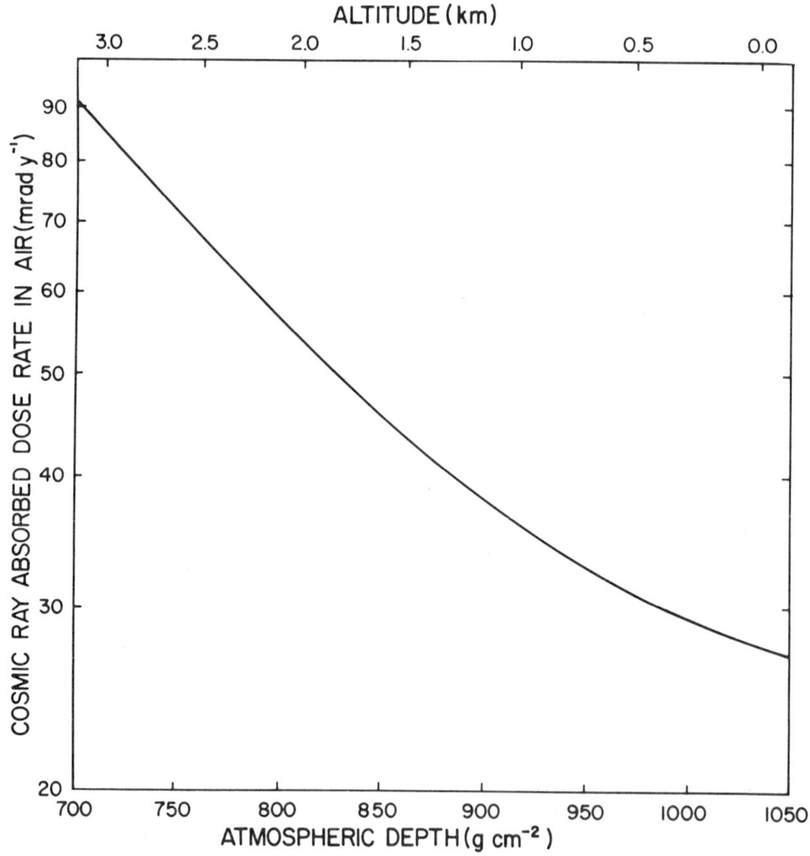

Fig. 2-2. Absorbed dose rates in free air from cosmic-ray charged particles in the lower atmosphere at geomagnetic latitude 50°N (from Lowder and Beck, 1966). (10 mrad y^{-1} = 1.14 μrad h^{-1}.)

in an equivalent thickness (g cm^{-2}) of less dense materials because of the reduced decay. However, denser materials more strongly attenuate the electrons produced by muon decay in air. These decay electrons are not replaced by electrons produced in the media and the total charged particle intensity and dose rate are therefore reduced approximately 30 percent by 50 g cm^{-2} of typical building materials and only by about 5 percent by each additional equivalent thickness. The attenuation by various types of buildings ranges from nearly zero for woodframe houses to more than 50 percent for lower floors in multi-storied office or apartment buildings. Mass thicknesses of the order of 10^4 g cm^{-2} are required to reduce the muon component to a few percent of its outdoor intensity.

As indicated in Table 2–10, cosmic-ray charged particles contribute

TABLE 2-13—*Amplitudes of cosmic-ray time variations in the earth's atmosphere*

Variation	Component	Atmospheric Depth	Latitude	$\Delta N/N_0 \times 100$[a]	Period	Reference[b]
		g cm^{-2}		percent		
Solar Modulation	Charged Particles	320	>50°	22	11 y	1, 2
	Charges Particles	500	>50°	16	11 y	1, 2
	Muons	1033	>50°	5	11 y	3
	Neutrons	680	>50°	25	11 y	4
	Neutrons	1033	>50°	20	11 y	4
Solar Flares	Muons	1033	>50°	0–400	h	5, 6
	Neutrons	1033	all	0–10000	h	6
Pressure	Charged Particles	1033	all	<10	d	
	Neutrons	1033	all	<15	d	
Temperature	Muons	1033	all	<5	1 y	7

[a] Total amplitude of variations relative to "normal" or average intensity (N_0).

[b] References: 1—Anderson (1961); 2—Neher (1967); 3—Forbush (1958); 4—Simpson and Wang (1970); 5—Dorman *et al.* (1956); 6—Sandstrom (1965); 7—Carmichael *et al.* (1967).

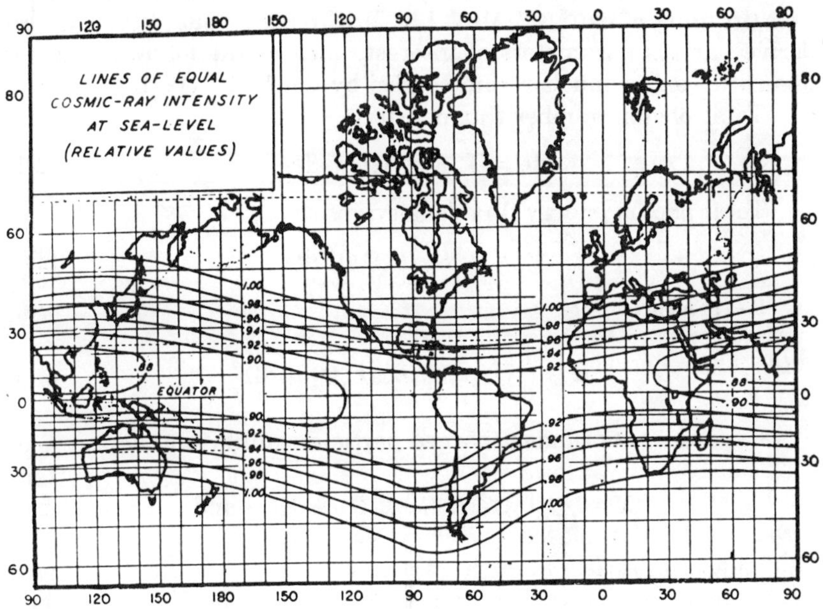

Fig. 2-3. Sea-level measurements of cosmic-ray ionization with shielded ionization chamber (1933–1935). At high geomagnetic latitudes the ionization rate is constant [from Millikan, R. A. and Neher, H. V. (1936). "A precision world survey of sea level cosmic-ray intensities," Phys. Rev. **50**, 15, by permission].

approximately 20 percent of the total free-air absorbed dose rate and 40 percent of the gonad dose rate from external radiation in typical situations. NCRP (1975) estimated the gonadal dose equivalent to the

U. S. population from cosmic-ray charged particles and neutrons to be about one-third of the total from all external and internal radiation, taking into account structural and body shielding and the altitude distribution of the population.

2.3.3. *Terrestrial Gamma Radiation*

2.3.3.1 *Physical Properties*. The gamma-radiation field depends on the radionuclide concentrations in various environmental media and the spatial distribution of such nuclides and the media relative to the measurement location. The gamma-ray response of a detector depends on the energy absorption properties of the detector, the incident photon number and energy flux density distributions. The absorbed dose to the detector is usually not closely related to the dosimetric quantity desired.

The most basic parameter characterizing a gamma-radiation field is the photon flux density, $\varphi(E, \Omega)$, as a function of energy and direction (in units of $cm^{-2}\ s^{-1}\ MeV^{-1}\ sr^{-1}$). From this quantity, one can derive various parameters of interest, such as the following, using quantities and symbols recommended by the ICRU (1971a).

Total photon number flux density:

$$\Phi = \iint \varphi(E, \Omega)\mathrm{d}E\,\mathrm{d}\Omega \tag{2-1}$$

Total photon energy flux density:

$$\Psi = \iint E\varphi(E,\Omega)\mathrm{d}E\mathrm{d}\Omega \tag{2-2}$$

Kerma rate:

$$\dot{K} \propto \frac{1}{\rho}\iint \mu_{tr}(E)E\varphi(E,\Omega)\mathrm{d}E\mathrm{d}\Omega \tag{2-3}$$

Exposure rate (under charged particle equilibrium):

$$\dot{X} \propto \frac{1}{W}\iint \mu_{en}(E)E\varphi(E,\Omega)\mathrm{d}E\mathrm{d}\Omega \quad \text{and} \tag{2-4}$$

Ionization rate:

$$I \propto \frac{\dot{D}}{\bar{W}} \tag{2-5}$$

where μ_{tr} and μ_{en} are the linear energy transfer and energy absorption coefficients, respectively, ρ is the density, $\dot{D}$ is the absorbed dose rate in free air, and $\bar{W}$ is the mean energy expended per ion pair formed. The doubly differential flux density can be folded in with the angular and energy response efficiencies of a detector to obtain the absolute detector response per unit total flux.

The absorbed dose rate, $\dot{D}$, in Eq. 2–5 is numerically equal to the kerma rate, and proportional to the exposure rate, under the condition of electronic equilibrium. This condition is closely approximated in free air in environmental gamma-ray and cosmic-ray fields. Thus, determinations of I and $\dot{X}$ for photons yield good estimates of absorbed dose rate in free air.

The unscattered photon flux density, $\varphi(E_0) \equiv \int \varphi(E_0,\Omega)d\Omega$, from a particular radionuclide distributed in the environment can be related to the source distribution by means of the general relation,

$$\varphi(E_0) = \frac{1}{4\pi}\int S(E_0, r) \cdot e^{-\int_0^r \mu_t(r)dr} \cdot r^{-2}\, dV \tag{2-6}$$

where S is the photon source strength at radius vector r, relative to the origin. For source and medium geometries, representing many environmental situations fairly accurately, analytical solutions to this equation can be obtained. Such geometries include infinite homogeneous, semi-infinite half-space, or infinite plane source distributions. In such cases, a measurement of $\varphi\ (E_0)$ yields specific values for $S\ (E_0,r)$.

Kogan *et al.* (1969) discuss methods for calculating the unscattered and total photon flux distributions for various source configurations of geophysical interest, including disk, sphere, rod, and strip sources. Semi-infinite, homogeneous half-space, and exponential distributions with depth have been treated in detail by Beck *et al.* (1972b) and Beck (1972b). These distributions are applicable to most situations of natural and fall-out gamma-ray emitters in the ground.

Unscattered and total photon flux densities and total exposure rates in air near the ground interface are related to source concentration of various exponential distributions defined by the parameter α in the relation:

$$S\ (z) = S(0)e^{-\alpha z} \tag{2-7}$$

where S is the photon emission rate per unit volume, z is the perpendicular distance in the soil from the soil-air interface, and α is the reciprocal of the "relaxation length" that represents the exponential distribution of the source concentration. As $\alpha \rightarrow 0$, the distribution approaches uniformity, while as $\alpha \rightarrow \infty$, it approaches a plane source at the interface. The former case usually applies to the principal natural emitters, while the latter case is approximately applicable to recently deposited fallout nuclides. Monte Carlo calculations of the doubly differential photon flux at one meter above a semi-infinite half-space for various monoenergetic sources and for the natural radionuclides have been published by Minato (1971).

The total primary photon flux density of energy E_0 at h cm above a flat air-ground interface from a particular radionuclide distributed exponentially with depth is

$$\varphi(E_0) = 2\pi \int_0^{\pi/2} \int_{h/w}^{\infty} \frac{S_0\,(E_0)}{4\pi r^2} \exp\,[-(\alpha/\rho)\rho z]r^2 \sin\theta \exp\,[-\mu_s(r - h/w) - \mu_a(h/w)]\; dr\; d\theta \qquad (2\text{-}8)$$

where w = cos θ,
S_0 = photon emission rate per unit volume as $z \rightarrow 0$ ($cm^{-3}s^{-1}$),
ρ = soil bulk density (g cm^{-3}), and
μ_a, μ_s = air and soil total linear attenuation coefficients, respectively (cm^{-1}).

Eq. 2–8 is a special case of Eq. 2–6 applicable to this source and medium geometry.

The dependence of $\varphi(E_0)$ on the angle of incidence (θ) with respect to the perpendicular to the earth-air interface ($\theta = 0$) is obtained from Eq. 2–8 by integration over r, i.e.,

$$\varphi(E_0, w) = \frac{S_0/\rho}{2} \frac{\exp(-t/w)}{(\alpha/\rho)w + (\mu_s/\rho)} \qquad (2\text{-}9)$$

where $t = \mu_a h$ (the detector height above the interface in units of mean free path). For natural emitters, $\alpha \approx 0$, and Eq. 2–9 becomes simplified, though the integration over w requires numerical methods with the aid of a large computer, even with this simplification.

Calculations of scattered photon flux density, angular and energy distribution of the total flux density, and exposure rate for exponentially distributed sources have been carried out by using a polynomial series-expansion approximation to the Boltzmann transport equation (Bennett and Beck, 1967). This technique accounts for the differences in photon transport in soil and air and the ±5-percent accuracy is verified by detailed experiments. The *in situ* soil composition, in weight percent, was 13.5 percent Al_2O_3, 4.5 percent Fe_2O_3, 67.5 percent SiO_2, 4.5 percent CO_2, and 10 percent H_2O, while the air density, 1.204 mg cm^{-3} was assumed constant with height between 0 and 100 meters. The distribution of source energies in the uranium and thorium series is given in Table 2–14, derived from the summary of individual radionuclide data in Tables 2–1 and 2–2. For the non-series radionuclides such as ^{40}K and ^{137}Cs, the source energy data are given in Tables 2–3 and 2–5.

The calculations of photon flux density provide guidance to interpreting *in situ* ground-level and airborne measurements, and labora-

TABLE 2-14—*Energy distribution of gamma rays emitted as a result of the decay of ^{238}U and ^{232}Th and daughters*[a]

Energy Interval	Number of Photons Emitted per Disintegration	
	^{238}U Series	^{232}Th Series
keV		
50–150[b]	0.139	0.062
150–250	0.104	0.555
250–350	0.196	0.277
350–450	0.361	0.021
450–550	0.022	0.156
550–650	0.436	0.313
650–750	0.027	0.081
750–850	0.084	0.139
850–950	0.032	0.351
950–1050	0.014	0.240
1050–1350	0.252	0.023
1350–1650	0.137	0.148
1650–1950	0.218	0.007
1950–2550	0.081	0.001
>2550	0.002	0.360
Total	2.1	2.7
Average series gamma energy	0.81 MeV	0.88 MeV

[a] From Beck (1972b).
[b] Omits x rays.

tory analysis of collected samples. Table 2-15 contains unscattered photon flux densities at a location 1 meter above the air-ground interface as a function of photon energy and source distribution. These data can be corrected to account for different source strengths, bulk densities and relaxation lengths (Beck *et al.*, 1972b) and are the basis for calibrating spectrometers (Section 4.5.1).

The integral energy spectra of the *total* flux density and *total* exposure rate at 1 meter and at 100 meters for the natural emitters plus ^{137}Cs are given in Table 2-16. These integral spectra indicate that the differential energy distribution of the photon flux density above the air-ground interface is not a sensitive function of the relative contributions by radionuclide sources. Photons with energies less than 140 keV contribute ~40 percent of the total number flux density at one meter height, but less than 10 percent of the exposure rate. As described in Section 4.2.2, this observation has an important bearing on the calibration of instruments, the responses of which are proportional to flux density rather than to absorbed dose rate.

Table 2-17 gives the exposure rate at one meter height per unit source for various gamma-ray energies and exponential depth distributions. Tables 2-18 and 2-19 give exposure rates per unit source for

TABLE 2-15—*Unscattered flux density at one meter above ground for distributed sources in the soil*[a, b]

Source Energy	Source distribution (α/ρ in cm^2/g^{-1})						
	0 (Uniform)	0.0625	0.206	0.312	0.625	6.25	∞ (Plane)
keV	$cm^{-2}s^{-1}$	$cm^{-2}s^{-1}$	$cm^{-2}s^{-1}$	$cm^{-2}s^{-1}$	$cm^{-2}s^{-1}$	$cm^{-2}s^{-1}$	$cm^{-2}s^{-1}$
50	1.44	0.082	0.225	0.305	0.475	1.147	1.577
100	2.77	0.146	0.363	0.471	0.679	1.359	1.710
150	3.33	0.170	0.410	0.526	0.744	1.427	1.775
200	3.91	0.184	0.455	0.577	0.802	1.483	1.804
250	4.06	0.201	0.470	0.591	0.819	1.506	1.863
364	4.72	0.227	0.516	0.643	0.878	1.578	1.933
500	5.39	0.252	0.560	0.692	0.9334	1.650	1.995
662	6.15	0.279	0.604	0.741	0.989	1.719	2.054
750	6.53	0.292	0.626	0.765	1.015	1.752	2.084
1000	7.53	0.325	0.677	0.821	1.077	1.830	2.151
1173	8.15	0.344	0.707	0.853	1.113	1.874	2.189
1250	8.44	0.352	0.720	0.868	1.129	1.895	2.205
1333	8.75	0.362	0.734	0.883	1.145	1.914	2.224
1460	9.15	0.373	0.751	0.901	1.166	1.941	2.247
1765	10.09	0.400	0.790	0.943	1.211	1.997	2.294
2004	10.82	0.419	0.817	0.973	1.243	2.036	2.334
2250	11.40	0.436	0.841	0.998	1.271	2.071	2.358
2500	12.17	0.454	0.867	1.025	1.300	2.105	2.385

[a] Flux densities are normalized to photon emission rates (Eq. 2.7) of $S(0) = \alpha$ cm^{-3} s^{-1} for $0 < \alpha < \infty$ and $S(z) = \rho$ cm^{-3} s^{-1} for $\alpha = 0$. For the special case of $\alpha = \infty$, a surface emission rate of 1 cm^{-2} s^{-1} is used.

[b] From Beck *et al.* (1972b).

uniformly-distributed natural radionuclides and for various depth distributions of fallout radionuclides.

Table 2-15 and 2-17 also illustrate the sensitivity of unscattered flux and total exposure rate to changes in soil density, particularly water content. The calculations are based on a water content of 10 percent by weight, which is reasonably typical for temperate regions. Varying the proportions of water from 0 to 25 percent by weight does not change the effective gamma attenuation coefficients and scattering cross sections by more than a few percent (Beck *et al.* 1972b). An increase in soil moisture reduces the source concentration per gram, and, for a uniformly-distributed source, the unscattered flux density and total exposure rate decrease proportionately. For exponentially distributed sources the net result is a decrease in α/ρ, the effect of which is evident from the tables.

Figure 2-4 shows the relative exposure rate contributions from various depths in soil for a uniform distribution of ^{40}K, the ^{232}Th series, and the ^{238}U series, calculated on the assumption that the relative contributions to the total exposure rate are 40, 40, and 20 percent, respectively. Figure 2-5 is a comparison of the energy distri-

TABLE 2-16—*The percentage of total flux density or total exposure rate from photons with energy less than E at 1 m and 100 m*[a]

	Energy (E in keV)									
	50	75	100	125	150	300	500	900	1500	2000
	percent	percent	percent	percent	percent	percent	percent	percent	percent	percent
Flux Density—1 m										
^{40}K	3	13	23	31	37	55	65	73	100	100
^{238}U series	4	17	30	39	45	67	79	89	96	99
^{232}Th series	4	17	27	37	44	67	75	86	92	94
^{137}Cs (α/ρ = 0.2)	3	13	22	31	35	55	65	100	100	100
Flux Density—100 m										
^{40}K	9	25	37	44	54	70	78	86	100	100
^{238}U series	10	29	42	50	61	78	86	92	97	99
^{232}Th series	10	30	43	52	59	76	84	92	95	96
^{137}Cs (α/ρ = 0.2)	10	29	42	50	61	78	86	100	100	100
Exposure Rate—1 m										
^{40}K	1	2	3	4	6	12	16	28	100	100
^{238}U series	1	3	6	8	11	24	38	56	78	96
^{232}Th series	1	3	5	6	9	20	28	45	60	67
^{137}Cs (α/ρ = 0.2)	1	3	4	5	8	18	30	100	100	100
Exposure Rate—100 m										
^{40}K	3	5	7	9	12	19	28	43	100	100
^{238}U series	3	7	12	14	18	33	46	65	80	97
^{232}Th series	3	6	9	11	14	27	36	50	63	71
^{137}Cs (α/ρ = 0.2)	5	10	14	17	22	40	58	100	100	100

[a] From Beck (1972b).

bution of the total flux density and the exposure rate from this type of a natural radiation source (Beck, 1972b). Most of the exposure rate is due to the small fraction of high energy photons from sources in the first few centimeters of soil. Therefore, the depth distribution data for environmental radionuclides for determining radiation levels need encompass only the top layers.

By employing published decay scheme data (Tables 2-1 and 2-2, or Martin and Blichert-Toft, 1970), Tables 2-15 and 2-17 can be used to obtain data for radionuclides not included in Tables 2-18 and 2-19. The data in Tables 2-15 and 2-17 also illustrate the strong dependence of primary flux density and total exposure rate on the depth distribution of the gamma-emitting nuclide in the ground. It is important to note, however, that the primary flux density per unit exposure rate (obtained by dividing the values given in Table 2-15 by the comparable ones in Table 2-17) is much less sensitive to the value of α. Thus, when the actual source distribution is poorly known, a primary flux density measurement still provides a reasonably accurate estimate of the air exposure rate contribution from a particular radionuclide, despite the uncertainty in the source distribution. This fact is

TABLE 2-17—*Calculated exposure rate at one meter above ground for distributed monoenergetic sources in the soil*[a,b]

Source Energy	Source distribution (α/ρ in cm² g⁻¹)						
	0 (Uniform)	0.0625	0.206	0.312	0.625	6.25	∞ (Plane)
keV	μR h^{-1}	μR h^{-1}	μR h^{-1}	μR h^{-1}	μR h^{-1}	μR h^{-1}	μR h^{-1}
50	0.88	—	—	—	—	—	—
100	2.05	~0.095	0.185	0.215	0.270	0.400	0.438
150	3.39	0.140	0.285	0.335	0.418	0.620	0.700
200	4.88	0.200	0.390	0.460	0.570	0.845	0.960
250	6.37	0.258	0.491	0.583	0.731	1.08	1.25
364	10.2	0.404	0.771	0.896	1.11	1.63	1.91
500	14.4	0.558	1.03	1.23	1.52	2.27	2.60
662	19.6	0.738	1.37	1.60	1.97	2.95	3.39
750	22.6	0.837	1.54	1.80	2.21	3.32	3.80
1000	30.4	1.10	2.00	2.32	2.85	4.28	4.86
1173	36.2	1.28	2.31	2.63	3.27	4.87	5.52
1250	38.4	1.33	2.41	2.79	3.42	5.14	5.86
1333	41.8	1.42	2.56	2.95	3.62	5.35	6.16
1460	45.1	1.54	2.75	3.18	3.88	5.73	6.56
1765	54.6	1.78	3.25	3.75	4.40	6.45	7.78
2004	62.2	2.07	3.60	4.13	5.00	7.15	8.20
2250	69.5	—	—	—	—	—	—
2500	77.2	—	—	—	—	—	—
2750	85.0	—	—	—	—	—	—

[a] Exposure rates are normalized to photon emission rates (Eq. 2.7) of $S(0) = \alpha$ cm^{-3} s^{-1} for $0 < \alpha < \infty$ and $S(z) = \rho$ cm^{-3} s^{-1} for $\alpha = 0$. For the special case of $\alpha = \infty$, a surface emission rate of 1 cm^{-2} s^{-1} is used.

[b] From Beck *et al.* (1927b).

important for the application of *in situ* gamma spectrometry to exposure rate determination (Section 4.5.2), particularly for radionuclides whose depth distribution cannot be reasonably estimated *a priori*.

The angular distribution of unscattered photons from sources distributed in the ground obtained by Beck *et al.* (1972b) reflects the effect of solid angle as well as the lack of effect of air as an absorber. The flux density distribution is nearly uniform over solid angle in the downward hemisphere, as is the exposure rate contribution (Minato, 1971). The azimuthal symmetry of the gamma-ray field around the vertical axis, which reflects the symmetry of the sources, suggests that the detector should also be symmetrical in order to simplify the evaluation of its angular response.

The altitude profile of flux density and exposure rate indicated in Table 2-20 is relevant to the use of airborne detectors in geophysical prospecting and large-scale radiation surveys. Of particular interest is the fact that the total flux density at 100 meters height per unit flux density at one meter is nearly the same for each of the natural

TABLE 2-18—*Calculated total exposure rate at one meter above ground for natural emitters uniformly distributed in the soil*[a]

Isotope	Exposure Rate/Radionuclide Concentration	
	μR h^{-1}/pCi g^{-1}	μR h^{-1}/indicated concentration
^{40}K	0.179	1.49 per percent K
^{226}Ra + daughters	1.80	0.61 per 0.358×10^{-6} μg g^{-1} Ra[b]
^{214}Pb	0.20	0.07 per 0.358×10^{-6} μg g^{-1} Ra[b]
^{214}Bi	1.60	0.54 per 0.358×10^{-6} μg g^{-1} Ra[b]
^{238}U + daughters	1.82	0.62 per μg g^{-1} ^{238}U
^{232}Th + daughters	2.82	0.31 per μg g^{-1} ^{232}Th
^{228}Ac	1.18	0.13 per μg g^{-1} ^{232}Th
^{208}Tl	1.36	0.15 per μg g^{-1} ^{232}Th
^{212}Bi	0.09	0.01 per μg g^{-1} ^{232}Th
^{212}Pb	0.09	0.01 per μg g^{-1} ^{232}Th

[a] From Beck *et al.* (1972b).
[b] Concentration of ^{226}Ra in equilibrium with 1 μg g^{-1} ^{238}U.

TABLE 2-19—*Calculated total exposure rate at one meter above ground for selected radionuclides distributed in the soil*[a]

Isotope	Source Activity	Source distribution (α/ρ in cm^2 g^{-1})					
		0.0625	0.206	0.312	0.625	6.25	(Plane)
	mCi km^{-2}	μR h^{-1}	μR h^{-1}	μR h^{-1}	μR h^{-1}	μR h^{-1}	μR h^{-1}
^{144}Ce	1.0	6.25(−5)[c]	1.34(−4)	1.56(−4)	1.96(−4)	2.86(−4)	3.27(−4)
^{144}Ce+^{144}Pr[b]	2.0	1.85(−4)	3.51(−4)	4.05(−4)	5.03(−4)	7.22(−4)	8.34(−4)
^{141}Ce	1.0	2.60(−4)	5.23(−4)	6.21(−4)	7.68(−4)	1.15(−3)	1.31(−3)
^{151}I	1.0	1.56(−3)	2.92(−3)	3.35(−3)	4.20(−3)	6.91(−3)	7.28(−3)
^{125}Sb	1.0	1.77(−3)	3.33(−3)	3.82(−3)	4.86(−3)	7.14(−3)	8.29(−3)
^{140}Ba	1.0	7.74(−4)	1.45(−3)	1.69(−3)	2.09(−3)	3.16(−3)	3.66(−3)
^{140}La	1.0	8.98(−3)	1.63(−2)	1.88(−2)	2.40(−2)	3.56(−2)	3.96(−2)
^{140}Ba+^{140}La[b]	2.15	1.11(−2)	2.02(−2)	2.33(−2)	2.97(−2)	4.40(−2)	4.92(−2)
^{103}Ru	1.0	1.97(−3)	3.66(−3)	4.30(−3)	5.37(−3)	7.90(−3)	9.22(−3)
^{106}Ru+^{106}Rh[b]	2.0	7.74(−4)	1.43(−3)	1.67(−3)	2.11(−3)	3.17(−3)	3.65(−3)
^{137}Cs	1.0	2.31(−3)	4.29(−3)	4.99(−3)	6.17(−3)	9.24(−3)	1.06(−2)
^{95}Zr	1.0	3.02(−3)	5.51(−3)	6.36(−3)	7.81(−3)	1.17(−2)	1.35(−2)
^{95}Nb	1.0	3.15(−3)	5.74(−3)	6.66(−3)	8.14(−3)	1.24(−2)	1.41(−2)
^{95}Zr−^{95}Nb	3.155	9.91(−3)	1.79(−2)	2.07(−2)	2.54(−2)	3.84(−2)	4.39(−2)
^{54}Mn	1.0	3.40(−3)	6.29(−3)	7.22(−3)	8.88(−3)	1.34(−2)	1.54(−2)
^{60}Co	1.0	9.99(−3)	1.80(−2)	2.06(−2)	2.55(−2)	3.78(−2)	4.32(−2)

[a] From Beck *et al.* (1972b).
[b] Assuming daughter is in equilibrium with parent; exposure rate is for 1 mCi km^{-2} of parent activity.
[c] Format of $6.25(-5) = 6.25 \times 10^{-5}$.

emitters uniformly distributed in the ground, and not too different for the main fallout radionuclide, ^{137}Cs. This suggests that a flux density measurement at 100 meters can be related to a one meter exposure rate fairly accurately. This has been demonstrated in several air-ground intercalibrations (e.g., Pensko *et al.*, 1971; Burson *et al.*, 1972).

The calculated results such as those in Tables 2-18 and 2-19 are slightly different from data previously reported by the same group (e.g., Beck and de Planque, 1968) and others as summarized by

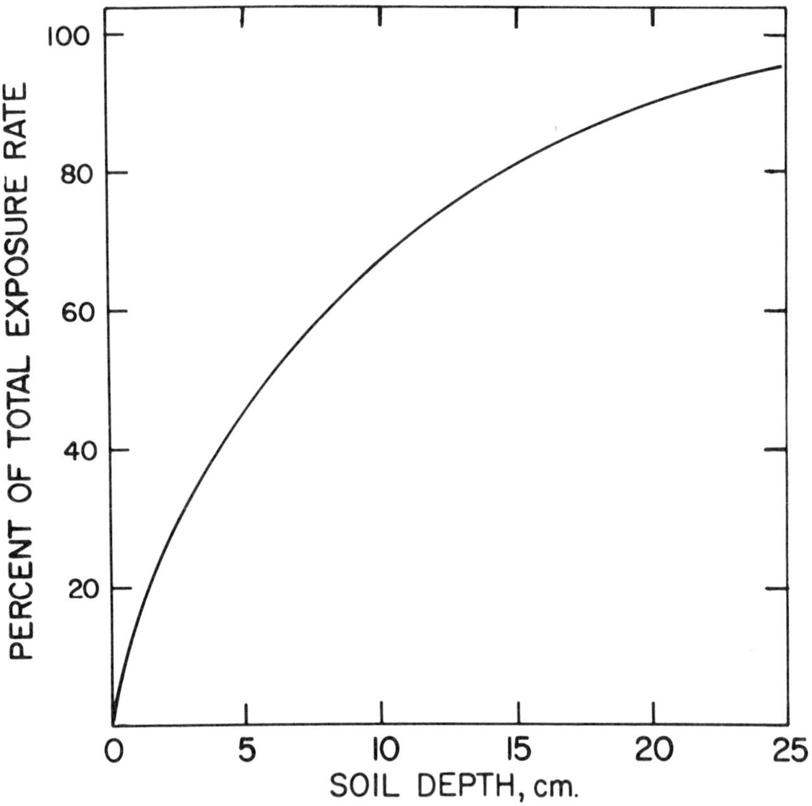

Fig. 2-4. Calculated relative contribution to the total exposure rate at one meter above the ground from natural sources as a function of soil thickness (from Beck, 1972b).

Kogan *et al.* (1969). These differences reflect both developing sophistication in transport calculations, and differing soil compositions (especially assumed water content), decay scheme data for the source radionuclides, and depth distributions for fallout. The calculations are strictly applicable only to the outdoor (air-ground) environment. The gamma radiation field within buildings would be expected to be more nearly isotropic and perhaps somewhat different in its energy distribution. However, for low-Z materials such as wood and ordinary concrete more than a few centimeters thick, outdoor data could be used for interpreting indoor measurements if proper account is taken of the differences in source-detector geometry. Unfortunately, in many practical cases, one does not encounter reasonably uniform media or source distributions in the indoor environment, and the gamma-ray field indoors cannot be treated with the generality possible for the outdoor environment.

2.3.3.2 *Measurements of Exposure Rate.* Exposure rate measurements have been made in various environments, both indoors and outdoors, in various parts of the world. The surveys have been summarized in the reports of the United Nations Scientific Committee on the Effects of Atomic Radiation, most recently in UNSCEAR (1972). These various measurements indicate that typical environmental gamma-radiation exposure rates range from 3 to 9 μR h^{-1} in both indoor and outdoor environments. In some regions, levels ap-

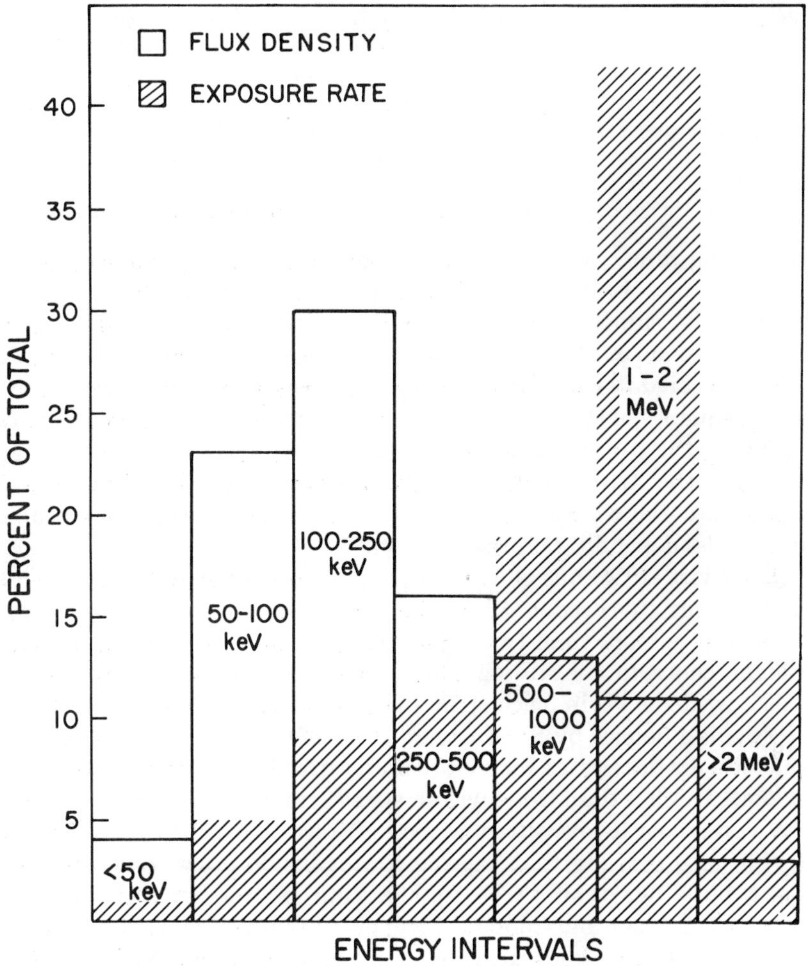

Fig. 2-5. Relative contributions to the total flux density and exposure rate at one meter above ground for photons of various energies for a uniform distribution of naturally-occurring sources in the ground (from Beck, 1972b).

TABLE 2-20—*Relative total photon flux density and total exposure rate for various heights above ground*[a]

	Height[b] (meters)					
	0	1	10	30	100	300
	percent	percent	percent	percent	percent	percent
Flux Density						
^{40}K	101	100	98	88	65	21
^{238}U series	101	100	93	83	60	14
^{232}Th series	101	100	93	82	59	16
^{137}Cs (α/ρ = 0.21)	103	100	92	79	49	12
Typical natural field	101	100	95	86	61	18
Exposure Rate						
^{40}K	102	100	89	74	44	12
^{238}U series	102	100	89	73	42	10
^{232}Th series	102	100	89	74	43	12
^{137}Cs (α/ρ = 0.21)	104	100	81	61	29	5
Typical natural field	102	100	89	74	44	12

[a] Reference: Beck (1972b).

[b] Values are normalized to 100 percent at 1 m to reflect common practice.

proaching zero have been observed, while in others, levels of over 1 mR h^{-1} have been recorded. Indoor gamma levels show approximately the same degrees of variation within regions and between them as do the outdoor levels, although the mean indoor value within a region can differ significantly from the mean outdoor value. This is a consequence of three factors: (a) the effect of structural shielding in reducing the contribution of outdoor sources relative to those originating in the structure itself, (b) the different uses of building materials, and (c) the wide range of radionuclide concentrations in building materials, depending on their origin.

Hultqvist (1956) has shown, in his survey of over 1600 apartments and houses in Sweden, that the gamma-ray ionization rate was consistently more than two times higher in structures built of lightweight concrete containing alum shale than those constructed of wood, while levels in brick structures lie between. The survey of 2026 Norwegian houses by Storruste *et al.* (1965) in several cities showed that the air dose rates inside brick and granite structures were consistently higher than those within structures built of other materials. Data on indoor levels, summarized by Yeates *et al.* (1972) and Oakley (1972), also indicate that exposure rates inside brick, concrete, and stone buildings tend to be higher than in wooden structures or outdoors. Lowder and Condon (1965) noted that the gamma-ray levels inside woodframe houses in urban areas of northern New England were consistently 70 percent of the outdoor levels. Thus, for some construction materials such as wood, the reduction of the ambient outdoor gamma-ray levels by attenuation, as well as by distance from the outdoor source, is greater than the contribution to the indoor radiation field from radionuclides contained in the material. The reverse is true for granite and some types of brick and concrete, though very wide variations have been observed in the radioactive contents of brick and concrete. This non-uniform composition of most structures makes the study of large-population exposure to gamma radiation by means of measurements of random samples difficult.

The individual contributions to the total gamma-ray dose rate from the important natural and fallout radionuclides have been determined by gamma-ray spectrometry (see Section 4.5). The major contributions to the normally encountered exposure rates are from ^{40}K, the uranium and thorium series, and ^{137}Cs from world-wide fallout (Lowder *et al.*, 1972), the thorium series and ^{40}K being the most significant contributors. Exposure rates from the uranium series are reduced from the values that would be observed under equilibrium conditions because of the escape of a considerable fraction of the ^{222}Rn from the upper soil layers to the atmosphere. Under many atmospheric conditions, the decay of the gamma-emitting daughters of ^{222}Rn in the atmosphere contributes a small fraction of the total gamma exposure rate at ground level (Table 2-10), which is a lower exposure rate than if the parent radon had been retained in the soil matrix (Beck *et al.*, 1966).

Individual radionuclide exposure rates have been associated with soil concentrations of particular radionuclides, using the conversion factors in Tables 2-18 and 2-19 (assuming a uniform distribution for the natural emitters and an α/ρ value of 0.21 cm^2 g^{-1} for ^{137}Cs). The uranium mean soil content inferred from *in situ* spectrometry is usually an underestimate, because of the unknown degree of radon escape to the atmosphere as a function of soil depth. Limited data indicate that an escape fraction of several tens of percent (Beck, 1972b) may apply to most field situations. It can be inferred, then, that the indicated lower limit is not in serious error. Inferred concentrations of ^{40}K and the thorium series are usually consistent with the data given in Table 2-7 and the available measurements of soil radioactivity summarized by Baranov and Morozova (1973).

Measurements by Lowder *et al.* (1972) and others have shown large differences in the environmental gamma radiation levels from place to place, even over short distances, in areas with complex soil geology or in urban areas where topsoils of different composition and origin have been juxtaposed. Similar variations are observed inside buildings of non-uniform construction. All of these spatial variations must be considered in the design of radiation surveys.

2.3.3.3 *Time Variations.* Variations in photon exposure rates with time at a particular location can be related primarily to changes in source-detector geometry. Two significant source changes are the emanation of radon gas from soil and building materials and its subsequent diffusion and transport, and the deposition of fallout radionuclides from weapons tests and their subsequent erosion, leaching, and decay. However, the two most important changes are due to variations in soil moisture content and in snow cover. Although relatively little work has been done in studying these various effects quantitatively, sufficient information is available to indicate their general nature and significance.

The exposure rate above the air-ground interface from radon daughters in both the ground and the air depends critically on several environmental factors in addition to the radium content of the ground, including the emanation coefficient, porosity, and moisture content of the soil, the barometric pressure, atmospheric stability conditions, and wind speed and direction. Analogous factors relating to building materials and indoor air dynamics influence the radon daughter contribution to the indoor radiation field. The complex interaction of these various factors, discussed in Section 2.2 in terms of variations of environmental radon levels, produces comparable variations in gamma radiation levels from the airborne radon daughters. Atmospheric ^{222}Rn in equilibrium with its daughters produces a gamma-ray exposure rate of about 0.15 μR h^{-1} at the air-ground interface for a typical ground-level concentration of 0.1 pCi l^{-1} (Gibson *et al.*, 1969). Given the range of values for radon and radon daughter concentrations observed in near-surface air (Section 2.2), one would expect the contribution to the ground-level gamma field from airborne ^{222}Rn daughters to vary between nearly zero and about 1 μR h^{-1}, with daily and seasonal patterns superimposed on irregular short-term fluctuations. The washout of these daughters from the air to the ground surface by precipitation produces increases in the exposure rate of several hours duration amounting to several μR h^{-1} at ground level, resulting from the transformation of a significant proportion of the atmospherically-distributed source to a plane source, which, in turn, is effectively closer to a detector located near the interface. These short-term increases have been observed in the uranium series peaks of gamma-ray spectra (Foote, 1964) and in total radiation measurements with ionization chambers (Burch *et al.*, 1964; Gibson *et al.*, 1969).

It is important to note that the variations due to source changes in the atmosphere are superimposed on those due to source changes in the ground. When only vertical radon transport is considered, an increase in the radon level in the atmosphere is associated with increased emanation from the ground, i.e., the effective removal of gamma-emitting nuclides from the upper soil layers. For a normally-encountered uranium concentration in soil (say 3 ppm) and fractions of ^{222}Rn produced that enters the soil air (several tens of percent), the effect of total removal of the ^{222}Rn in the air spaces in the upper soil layers on the gamma radiation field would be a decrease in exposure rate of several tenths of μR h^{-1} one meter above the interface. This change is comparable and opposite to the change due to atmospheric effects.

The actual dynamics of radon distribution in the vertical direction at an outdoor measurement site are complicated by the horizontal atmospheric transport to the site of radon produced elsewhere. This effect may be the main cause of significant changes in the ^{222}Rn daughter contribution to the ground level gamma field at many locations.

The above considerations are of little significance for ^{220}Rn, the relatively short half-life (54.5 sec) of which precludes it from being transported far from its point of production. The assumption that all of the gamma-emitting daughters of ^{220}Rn in the thorium series are retained in the soil with a uniform distribution is generally satisfactory in interpretations of gamma-ray spectra in terms of air exposure rate (Beck *et al.*, 1966; 1972b), although some do escape the ground. However, small increases in ^{220}Rn daughter gamma radiation have been noted in field gamma spectra after precipitation (Foote, 1964).

Changes in the moisture content of soil in effect modify the radionuclide depth distribution. When moisture fills the air spaces in soil, it has the effect of increasing bulk soil density without significantly affecting its gamma-ray attenuation properties. Thus, the photon flux density and exposure rate above the ground change in proportion to the source concentration. For exponentially distributed sources such as fallout, the effect is to change the value of α/ρ (Table 2-19). For the uniformly-distributed uranium series, the effect of soil density change is superimposed on the effect of change in soil porosity due to moisture in the pore spaces, which influences radon transport. The complexity of these phenomena is increased by the fact that the moisture content of the soil may be more nearly exponential with depth than uniform, particularly within a day after significant rainfall (Kol', 1952; Kogan and Fridman, 1967). In any case, Koga- and Fridman (1967) estimate that changes in the gamma radiation field due to seasonal variations in soil moisture are of the order of 10 percent and larger decreases are possible for periods of up to several days after a heavy rainfall. Such changes have been confirmed in long-term gamma radiation monitoring experiments with ionization chambers (e.g., Thompson and Wiberg, 1963; Burch *et al.*, 1964; Gibson *et al.*, 1969), gamma spectrometers (Foote, 1964), and thermoluminescence dosimeters (Burke, 1975). The spectrometric data of Foote (1964) also illustrate the difference in the net effect of precipitation on the gamma-ray field from the uranium series as compared with that from potassium and the thorium series. On several occasions, reduction in the latter of 15 percent was accompanied by no net change in the contribution from the uranium series.

Snow or standing water acts as an absorber of radiation from the ground, the extent of such shielding depending on the snow water content. Snow also introduces a strong seasonal pattern on outdoor environmental gamma-radiation exposure rates (Thompson and Wiberg, 1963; Burch *et al.*, 1964; Pensko 1967; Gibson *et al.*, 1969). In addition, melting snow adds to the soil moisture content and produces source dilution in the soil matrix.

The contribution to the gamma radiation field in air from fallout radionuclides at any location exhibits a time dependence that is a function of deposition and radioactive decay rates and the rates of erosion from and leaching into the ground. Data on the significant changes in the total outdoor gamma radiation field during and shortly after periods of large-scale weapons tests in the atmosphere have been given by Thompson and Wiberg (1963); Burch *et al.* (1964); Beck (1966); Pensko (1967); and Gibson *et al.* (1969). Peak values for fallout gamma exposure rates comparable to typical natural gamma levels were attained in 1963, primarily due to ^{95}Zr-^{95}Nb. These levels declined rapidly during 1964–65. In recent years, the fallout gamma levels in the environment have remained nearly constant, usually between 0.1 and 1 $\mu R\ h^{-1}$ in the United States, with ^{137}Cs as the main contributing radionuclide (Lowder *et al.*, 1972). As most of the ^{137}Cs in the ground has been present for some time, changes in the ground distribution and consequent time variations of flux and exposure rate due to source deposition and migration are small and are of the order of 0.1 $\mu R\ h^{-1}$ over a given year. The seasonal pattern of deposition that includes input of both ^{137}Cs and shorter-lived fallout radionuclides from recent tests is now the main source of such variations.

The development of local vegetation is another possible source of seasonal change in environmental gamma-radiation levels. Perturbations in both source distribution and in the attenuating media may change the measured gamma-ray field by roughly 10 percent in forested areas (Kogan *et al.*, 1969).

The increasing use of radionuclides and nuclear power fuel cycles introduces new local sources of gamma radiation into the environment which may be strongly time-dependent. For example, the gamma exposure rate from the noble gases emitted from the stack of a light-water power reactor depends on the stack emission rate and the spatial orientation of the plume relative to the site, that in turn depends on the local meteorology. Under such conditions, pulses in gamma radiation intensity are observed that are similar to but more rapidly variable than those produced by radon-daughter fallout during precipitation (Lowder *et al.*, 1972). Such time-dependent phenomena can be used for discriminating between such manmade sources and the natural background levels (Beck *et al.*, 1971; 1972a).

Evidently, long-term measurements of gamma-radiation flux density and exposure rate at any point in the environment can be useful in investigations of interacting environmental factors, and permit a more adequate interpretation of exposure rate measurements in terms of long-term exposure.

2.3.4 *Terrestrial Beta Radiation*

Many of the important environmental radionuclides are beta emitters, most notably ^{40}K, ^{87}Rb, ^{90}Sr-^{90}Y, ^{137}Cs, ^{214}Pb and ^{214}Bi. The last two nuclides are decay products of ^{222}Rn, and now contribute the bulk of atmospheric beta radiation (see Table 2-8). By the 1980s, average concentrations of ^{85}Kr from nuclear power are expected to about equal those of the radon daughters. The contribution of ^{220}Rn daughters is of the order of 1 percent that of the ^{222}Rn daughters.

The range of 1 MeV beta particles in air and soil is approximately 0.4 g cm^{-2}, corresponding to 3 m of air and about 0.3 cm of soil. Therefore, the beta radiation field near the air-ground interface depends on the content of beta-emitting nuclides in the surface layer of soil and in the atmosphere within a few meters of the interface. The beta particle flux decreases rapidly above the interface to a level characteristic of the atmospheric radioactivity content.

The earliest measurements of beta particle ionization in the environment were carried out by Hess and his colleagues (Hess and O'Donnell, 1951; Hess *et al.*, 1953; Miranda, 1958). More recent measurements have been reported by Kawano *et al.* (1969); Ikebe (1970); and Iida and Kawano (1972). At one meter above the ground, all measurements indicate absorbed dose rates in free air between 1.0 and 2.5 μrad h^{-1}, except those made during and shortly after the periods of weapons testing in the atmosphere when they were an order of magnitude higher (Gibson *et al.*, 1969; Kawano *et al.*, 1969). By 1968, the levels had declined to the typical range indicated above, which still includes a small contribution from long-lived fallout emitters in the topmost layers of soil.

Measurements of the variations with height of the beta absorbed dose rate in air are consistent with the calculated profiles of O'Brien *et al.* (1958), and Iida and Kawano (1972). Within 10 cm of the ground surface, the levels can be in excess of 10 μrad h^{-1} and are comparable to the normal gamma absorbed dose rates at about 50 cm height.

The contribution of atmospheric radionuclides to the measured total beta absorbed dose rate in air is typically about 20 percent at one meter height, averaging about 0.4 μrad h^{-1}. However, given the usual range of ^{222}Rn levels in ground-level air (see Table 2-8), this contribution can vary from much less than 0.1 to about 1 μrad h^{-1}. Gibson *et al.* (1969) calculate a value of 0.82 μrad h^{-1} per pCi l^{-1} of ^{222}Rn in equilibrium with its daughters.

The time-varying water content of the upper soil layers and the effect of standing water or snow on top of the ground can be expected to produce strong variations in the beta-radiation field in air produced by ground sources. These variations have not yet been documented in detail, but may amount to an order of magnitude.

Because of their low concentrations in the lower atmosphere ($<$1 pCi m^{-3}) and the short range (a few cm in air) of their low energy beta particles, the cosmogenic radionuclides ^{3}H and ^{14}C make a negligible contribution to the terrestrial beta-radiation field.

2.3.5 *Terrestrial Alpha Radiation*

An important contribution to total ionization and absorbed dose rate in free air near the air-ground interface is produced by alpha particles emitted by ^{222}Rn, ^{220}Rn, and their daughters. Because of the short range of these particles in air ($\sim$6 cm), this ionization contribution is strongly dependent on the local air concentration of these nuclides and is thus highly variable. Special techniques must also be devised to measure the alpha ionization, as most radiation detectors have walls sufficiently thick to exclude the incident alpha particles.

Measurement of alpha-particle ionization by Hess and Vancour (1950); Kawano *et al.* (1965); Grevet (1966); and Gibson *et al.* (1969) indicate typical alpha absorbed dose rates in ground level air of 1 to 5 μrad h^{-1} and a range of 0.1 to more than 10 μrad h^{-1}. The time and altitude variations of this quantity would correspond to those of radon and its daughters. According to Ikebe (1970), an air content of 0.1 pCi l^{-1} of ^{222}Rn in equilibrium with its daughters (through ^{214}Bi) yields an alpha dose rate in air of 3.2 μrad h^{-1}. The corresponding figure for the same content of ^{220}Rn in equilibrium with ^{216}Po is 1.5 μrad h^{-1}.

2.3.6 *Terrestrial Neutrons*

Neutrons may be produced in terrestrial media by the spontaneous fission of heavy natural radionuclides such as ^{238}U, by cosmic-ray interactions, and by (α,n) or possibly (γ,n) reactions. UNSCEAR (1966) has estimated the neutron production rate in typical soils to be approximately 2000 g^{-1} y^{-1}. Of this, only about 20 g^{-1} y^{-1} would derive from fission or nuclear reactions from terrestrial radiation, adding less than 1 percent to the cosmic-ray neutron flux density at the earth-air interface.

3. Requirements for Measurement and Surveillance Programs

3.1 Rationale for Measurements

3.1.1 *Introduction*

Environmental radiation measurement programs are conducted for the various purposes indicated in Section 1 and in the literature, e.g., ICRP (1965a). It is important to consider the relation of measurements to dose assessment, especially in the monitoring of radionuclides which may enter the environment from nuclear facilities.

The assessment of dose to individuals or populations from particular sources of radiation depends on the nature of the source, its distribution in the environment, the type of radiation involved, and the significance of the estimated doses relative to other sources of exposure. Research indicates how radionuclides are distributed in and transported through the environment, and how the resulting radiation exposure of man can be modeled. Computational models are a necessary element in any dose assessment program, since measurements rarely lead to dose estimates directly and unambiguously.

The rationale for the assessment of dose from nuclear facility effluents is straightforward, although it involves knowledge of their radionuclide composition, the manner and extent of the radionuclide distribution in the environment, and possibly the characterization of nuclides from other sources.

Environmental monitoring intended for developing data for dose assessment should be based on careful consideration of the important radionuclide pathways through the environment (ICRP, 1965a). While it is important to avoid the accumulation of extraneous measurements, it is likewise important to assure that significant dose contributions are not neglected. Evaluation of the factors that influence the dose to man is a prerequisite to the selection of appropriate measurement methods. One often relies on more extensive measurements for developing or improving dose calculations.

In this report, a measurement method is the entire process from measurement or sample site selection through data interpretation. This process results in the determination of some physical quantity, which is an input into a dose assessment procedure. The dose inferred may, therefore, be subject to large errors, so each step of the measurement process and dose assessment should be taken carefully and tested. Suitable precautions for achieving adequate measurements are given in the following three sections.

Dose assessment may be conducted on a local, regional, national, continental, or global basis in several ways. Dose to man may be calculated from assumed or measured releases of radionuclides from known release points, taking account of their environmental dispersion, subsequent accumulation in soil, sediment or biota, and the manner and degree of exposure to man. This approach is used to provide knowledge about possible radiation exposure and guidance for the design of measurement or surveillance programs.

Alternatively, dose may be inferred from measurements of the radiation field or radionuclide concentrations in environmental media. For example, from the radionuclide concentration in a particular item of food and known consumption rates, the intake of the radionuclide is estimated and, from metabolic data for man, the dose associated with this intake is inferred.

These theoretical and empirical approaches are complementary. The first can be made before a facility begins operation or an activity undertaken using data and information from prior studies conducted either locally or around the world.

The United Nations Scientific Committee on the Effects of Atomic Radiation has published summaries of the world's literature regarding environmental radionuclides from nuclear weapons testing, most recently UNSCEAR (1972). These reports contain information about the transfer of radionuclides between the atmospheres, waters, soils, biota, and man and about his exposure to the various nuclides. Information on the dosimetric implications of the disposal of radioactive wastes into the sea has been published in IAEA (1961) and NAS-NRC (1959, 1962, 1971) and Polikarpov (1966), of the dispersion of radionuclides into rivers, lakes and estuaries in IAEA (1971a), and of radionuclide content of food in NAS-NRC (1973). The ecological behavior of many nuclides has been reviewed in NAS-NRC (1972) and discussed in symposia such as IAEA (1966a; 1969a); Schultz and Klement (1963); Nelson and Evans (1969); Nelson (1971); and Aberg and Hungate (1967).

The nature and extent of measurements for dose assessment are affected by whether or not one is primarily concerned with the few most exposed individuals or with the integrated exposure of population groups. In either case, account must be taken of living habits, including the fraction of time spent indoors.

The role of measurements of environmental radiation levels and radionuclide concentrations in dose assessment is thus complex and depends on the particular situation being studied. However, the description of common features of dose assessment aid in elucidating the role of measurement, the interplay of experiment and calculation, and the concept of dose pathway, thus providing guidance to designing measurement programs.

3.1.2 *Sample Collection Considerations*

The primary purposes of *in situ* gamma-ray measurements are to determine the radiation field properties, to identify radionuclides, and to determine their approximate concentrations. In most circumstances, however, collecting samples for laboratory radionuclide analysis is the primary approach. Sampling and analysis yields information on the identity and concentration of radionuclides, and can be used to describe their spatial distribution and temporal fluctuations. Laboratory analyses of environmental samples also provide data on the chemical and physical forms of the radionuclides, the source and origin of a release, and natural or other possibly important radionuclides. Long-term sampling is useful in determining inventories and trends.

General guidance for establishing radiological monitoring programs has been provided by several national and international groups (ICRP, 1965a; IAEA, 1966b; FAO, 1962; EPA, 1972), and symposia proceedings (for example, Godebold and Jones, 1965; Reinig, 1970; Voilleque and Baldwin, 1971; IAEA, 1971b).

The requirements for sampling include selecting the proper sampling medium, defining the sample, assuring that the sample is representative and maintains its form until analysis, applying analytical quality assurance procedures, and obtaining appropriate background or base-line values, (APHA, 1971; Black, 1965; Horwitz, 1965). A sampling program is best done by specialists in the various disciplines. Procedures for obtaining samples should always be tested to assure their reliability. In sampling for radionuclides, an important additional consideration may be their usually extremely low concentration. Very large samples must sometimes be collected and analyzed and very sensitive radiation detection instruments utilized. Care should be taken in analyzing such samples, since trace nuclides may differ in their behavior from the larger quantities of related stable nuclides during collection, storage, and analysis.

The decision on the extent of a sampling effort requires balancing needs with limitations on skilled persons, equipment, and funds for processing and analyzing collected samples. The astute planner limits analyses to the radionuclide samples and pathways that are the most important or critical ones, or to samples from the vicinity of individual persons or population groups.

Although discussions of statistical considerations for sampling designs are available (for example, Cochran, 1963; Wilson, 1952), their guidance is general. Criteria for selecting sampling locations surrounding a source of airborne radionuclides (Pelletier, 1970) and in a large contaminated field (Eberhardt and Gilbert, 1972) are based on such considerations. The design of a sampling program near a local source, as opposed to a global program, should account for the effects of local meteorological and terrestrial conditions and avoid the collection and analysis of large numbers of samples.

An *in situ* radiation survey should be undertaken to identify important media and locations. An alternate approach is to collect many samples, analyze a few of them, and store the rest. If several samples can be analyzed simultaneously without losing their identity (e.g., by gamma-ray spectrometry), relatively uncontaminated samples can be set aside with minimal effort. The sampling program can be contracted or expanded in proportion to the expected seriousness of the radiation exposure (Philbin and Whipple, 1970), with the more accurate determinations made for the larger radionuclide concentrations (ICRP, 1965a).

Samples that yield information most directly on radiation exposures of populations, air, food, and water, are usually of greatest interest in facilities monitoring. Radionuclide concentrations of concern have been published (ICRP, 1959; NCRP, 1959) together with values of air, food, and water intake by the "standard" adult. Dietary constitutents have been quantified in several reports (e.g., Hardy, 1974; Honstead, 1971; Bureau of the Census, 1963). The available analytical methods must be considered to select the most effective combination of sampling, chemical separation, and radionuclide measurement, i.e., one that yields the required information with least effort. Additional guidance on sample collection is given in Sections 5.1 and 5.2.

3.2 Pathway Analysis

3.2.1 *Features of Dose Assessment*

The general elements of dose assessment are:

(a) The characterization of the radionuclide or radiation source;

(b) The determination of radionuclide distributions;

(c) The determination of radiation incident on and/or the radionuclide accumulation by man;

(d) the determination of the consequent radiation dose to man.

These features are illustrated by considering the emission of particular radionuclides from a nuclear facility. Table 3-1 shows the various steps in the inference of the dose from radionuclides deposited in man following transport in the terrestrial environment. The activity released per unit time to the environment is dispersed in the air, at

TABLE 3-1—*Illustrative equilibrium pathway for internal radiation exposure from airborne radionuclide effluent*

Pathway Step	Variable	Product of Variables
Source	Source term (Q)	Radioactive material released per unit time Q
Dispersion	Dispersion (χ/Q)	Radioactive material per unit volume $Q(\chi/Q)$
Deposition	Deposition velocity (V_d)	Radioactive material per unit area and unit time $V_d(\chi/Q)Q$
Cumulative Deposition	Accumulation time (τ)	Radioactive material per unit area $\tau V_d(\chi/Q)Q$
Environmental and Biological Transfer[a]	Transfer coefficient (T_r)	Radioactive material per unit mass of environmental or biological medium $T_r\tau V_d(\chi/Q)Q$
Human Intake	Intake rate (I)	Radioactive material intake per unit time $IT_r\tau V_d(\chi/Q)Q$
Human Absorbed Dose Rate	Intake rate to absorbed dose rate conversion factor (k)	Absorbed dose per unit time $kIT_r\tau V_d(\chi/Q)Q$

[a] T_r is simplified here; T_r is the product of a number of coefficients, e.g. soil to air, soil to grass, grass to animal, animal to animal, etc., and depends on time and on metabolic, physical and chemical variables.

least partially deposited on soil or vegetation, and partially taken in by man via ingestion.

Dose assessment is performed both prior to and during facility operation. Dose assessment prior to operation is done by modeling the potential nuclide pathways and employing assumed facility radionuclide release data. In this type of assessment it is useful to normalize the dose estimate to the unit quantity of each important nuclide so that comparison with subsequent measurements is convenient. Subsequent measurements are restricted to those deemed critical.

In the following discussion, the pathways of practical interest in nuclear facilities monitoring of liquid and airborne effluents are emphasized. For pathways which might be critical, consideration is given to those points in the pathways that present opportunities for optimum measurement.

3.2.2 *External Irradiation*

3.2.2.1 *Source → Man*. Man can be exposed to radiation sources located at the nuclear facility and in pathways to man. The most significant radiation sources are generally photon emitters, although betas may be important, e.g., from ^{85}Kr near a reactor fuel reprocessing plant (Smith *et al.*, 1970). Although primarily a local shielding problem, an example of the direct radiation pathway is the exposure to the 6.13-MeV gamma radiation from ^{16}N in the turbines of a boiling water reactor (Lowder *et al.*, 1973) and gamma radiation from temporarily-stored waste material (Kahn *et al.*, 1971). Leakage and scattered radiation, largely high-energy neutrons, from large particle accelerators may also require evaluation and monitoring. Stephens and Cantelow (1975), for example, reported annual neutron dose-equivalent values ranging from several to about 13 mrem at selected monitoring locations near the site boundary. The proper evaluation of this type of pathway requires measurements of both the absorbed dose rate in free air and/or particle flux density as a function of particle type and energy.

3.2.2.2 *Source → Water → Man*. Radionuclides released into the environment as dissolved or suspended material in a liquid effluent can cause direct radiation exposure of individuals who are either immersed in the water or near the water surface. Again, photon emitters are of primary interest, although in the former case the possibility of a significant skin dose from betas should be considered. Dose rates are estimated either by radiation measurement or by calculation that uses the radionuclide concentration in the water. An example of the contribution to whole-body dose of this pathway relative to others has been given by Honstead (1971) for a rural population.

3.2.2.3 *Source → Water → Sediments → Man*. This pathway differs from the previous one in that many radionuclides are preferentially deposited on exposed sediments on the shores of streams, rivers, and lakes or on tidally-washed estuaries or coastal areas. Examples of this pathway are the accumulation of ^{106}Ru and ^{95}Zr-^{95}Nb on sediments in the coastal area of Britain (UNSCEAR, 1972) and of ^{134}Cs and ^{137}Cs on Hudson River sediments near a pressurized water reactor (Wrenn *et al.*, 1974). In both cases the dose contributions were comparable to those from other pathways, particularly the ingestion of fish and biota.

3.2.2.4 *Source → Air → Man*. External beta and gamma irradiation occurs upon immersion in air containing radioactive gases and aerosols, and only gamma irradiation occurs if a cloud or plume is located nearby. The determination of the integrated dose from such plumes by radiation measurements may require a long-term monitoring program because of the complex relationship between plume geometry and local meteorology. This pathway is significant for boiling water reactors with short holdup times (less than several hours) in the release of their gaseous effluent (Blanchard *et al.*, 1971) and for nuclear fuel reprocessing plants (Smith *et al.*, 1970).

3.2.2.5 *Source → Air → Soil → Man*. Radionuclides released to the atmosphere may be deposited on the ground and add significantly to the gamma radiation field near the air-ground interface produced by naturally-occurring sources. This is a major pathway for fallout from nuclear weapons tests (UNSCEAR, 1972), if the dose commitment to the year 2000 from tests made prior to 1971 is considered. Either direct radiation measurement or a determination of the nuclide distribution with soil depth may be used for dose assessment.

3.2.3 *Internal Irradiation*

3.2.3.1 *Source → Water → Man*. This pathway may be important, if the local water is consumed in large quantities by people of all ages. Particular attention should be paid to the removal of radionuclides by water treatment plants. Measurements of radionuclide concentrations in drinking water supplies should be carried out after such treatment. Examples of this pathway involving radionuclides is the Columbia River (Honstead, 1969; 1971) and possibly sites near nuclear power reactors if the liquid effluent may reach drinking water supplies.

3.2.3.2 *Source → Water → Algae → Man*. Man sometimes consumes food prepared from edible seaweed, so this pathway may be important whenever radionuclides are discharged into bodies of water where such algae grow. An example of this complex pathway is the Windscale liquid discharge, where the critical pathway was determined to be due to laverbread consumption by a population group (2×10^4 people) in South Wales (UNSCEAR, 1972).

3.2.3.3 *Source → Water → Shellfish → Man*. The consumption of shellfish represents a route of potential radiation exposure whenever releases of radionuclides are made in the vicinity of shellfish farms, though concentration factors for shellfish are generally less than those for algae. An example of this pathway is found near the Bradwell Nuclear Plant in Great Britain, where the consumption of oysters containing ^{65}Zn is the critical pathway (UNSCEAR, 1972).

3.2.3.4 *Source → Water → Fish → Man*. This may be the critical dose pathway in fresh water or whenever shellfish or algae in a marine or estuarine area are not consumed. An example is the Hudson River estuary where the consumption of fish containing ^{134}Cs and ^{137}Cs from a nearby pressurized water power reactor is a critical pathway (Lentsch *et al.*, 1971; Wrenn *et al.*, 1974). In this estuary, the water is too saline for drinking purposes and shellfish and algae are not used for human consumption.

3.2.3.5 *Source → Water → Soil → Food → Man*. This pathway is important whenever releases of radionuclides are made to bodies of water used for irrigation, as in the case of the uptake of radionuclides by crops irrigated by Columbia River water (Honstead, 1969; 1971). Radionuclides in water may also enter into man's food supply by means of uptake by animals foraging at the water's edge.

3.2.3.6 *Source → Air → Man*. The inhalation route of radionuclides released to the atmosphere directly from the source or resuspended after deposition may result in their absorption from the lung, gut, or skin. Absorption and subsequent distribution in the body depends on the physical state of the radionuclide. For example, inhaled radioiodine is absorbed into the blood and subsequently concentrated in the thyroid, while relatively insoluble nuclides can remain deposited in the lung and lymph nodes or subsequently translocated internally. Some airborne radionuclides may enter the body by transpiration through the skin.

3.2.3.7 *Source → Air → (Soil →) Vegetation → Man*. Airborne radionuclides attached to particulates are deposited on the surfaces of vegetation by dry convective deposition and by precipitation. The ingestion of vegetables and fruits with such surface deposition is generally not an important dose pathway to man because of the short residence time. An exception would be a high deposition rate of radioiodine on rapidly harvested leafy vegetables. In the case of absorption of radionuclides into the plant system from the leaves or through the roots, the uptake from the soil is dependent on the chemical and biological properties of the nuclide and of the soil. Radionuclides accumulated in the plant are available for direct ingestion by man or for ingestion by grazing animals.

3.2.3.8 *Source → Air → (Soil →) Vegetation → Meat or Milk → Man*. This pathway is frequently important, notably for releases of ^{131}I at nuclear facilities and for ^{90}Sr and ^{137}Cs in fallout from weapons tests. The bulk of the dose from these internally-deposited radionuclides may be derived from milk and/or meat in the diet. Although this is a complex pathway, a great deal of information about the various parameters and how radionuclides are transferred between the various media is available (e.g., UNSCEAR, 1972). Measurements of radionuclide concentrations in one medium are then related to those in other media and ultimately to absorbed dose via animal and then human ingestion.

3.3 Measurement Methodologies

3.3.1 *Dose From Internally-Deposited Radionuclides*

A common feature of the internal radiation pathways indicated in Section 3.2.3 is the transfer of radionuclides from medium to medium until they are inhaled or ingested by man. Research studies have indicated the degree to which this transfer takes place for certain radionuclides in particular pathways. Eisenbud (1973) and Pertsov (1973) have recently reviewed this subject area, and additional considerations have been reported by Klechkovskii *et al.* (1973).

Once the intake of a radionuclide has been estimated, the dose to a particular organ can be determined from models of radionuclide distribution in the body (e.g., ICRP, 1959) and of organ dose for particular distribution (Snyder *et al.*, 1969). Values for the important parameters in internal dose calculations have been given by UNSCEAR (1972) and Eisenbud (1973). The essential role of a measurement in the analysis of a dose pathway is to provide an observation in the environment from which the identity of the responsible nuclides and the magnitude of the dose may be determined. Because of the unavoidable uncertainties in the pathway model that include transfer and concentration factors in various media, dietary analyses, and calculations of organ dose per unit intake, the reliability of dose estimates is best for measurements of radionuclide content in media in the pathway close to man or in man himself.

A "classical" example of a complex dose pathway to man where measurements of radionuclide concentration in various media are possible is that due to the release of ^{131}I, its transport through the atmosphere and deposition on pasture land, its uptake by grazing animals and appearance in milk, and the consumption of the milk by man. For continuous releases, a quasi-equilibrium condition normally arises whereby the ^{131}I concentration in one medium can be unambiguously related to the source term, to concentration in other media, and to dose. For example, the absorbed dose rate $\dot{D}$, to the thyroid can be represented as a function of the ^{131}I activity in milk, C, in the following manner:

$$\dot{D} = \mathrm{k}IC \tag{3-1}$$

where I is the intake rate of fresh milk of local origin in the diet and k is the calculated dose rate to the thyroid for a constant intake rate of unit activity per unit time. The value k may vary with age, body weight, or sex, and I and C may fluctuate significantly with time. However, the desired dose is usually that over an extended period, for example one year, so that determination of appropriately time-averaged values of I and C is sufficient. Monitoring of radioiodine in milk would then be designed to obtain such an average value of C, and dietary studies of typical fresh milk intake rates would yield a value for I. Studies have shown that the value for k appropriate to a critical population, infants, (in the sense intended by ICRP, 1965a), is 16 μrad pCi^{-1} (UNSCEAR, 1972). For a fresh milk intake rate 1 l d^{-1} (averaged over a year) and a mean ^{131}I concentration of 1 pCi l^{-1}, the annual thyroid dose to these infants would be about 6 mrad. For adults, the dose would be much lower because of their generally smaller intake of fresh milk and the larger adult thyroid.

The ^{131}I concentration in milk can also be related to earlier stages of the pathway (see Table 3-1) by means of the equation

$$C = Q\,(\chi/Q)\,V_{\mathrm{d}}\,\tau\,T_{\mathrm{r}} \tag{3-2}$$

where Q is the source release rate (μCi s^{-1}), χ/Q is the average dilution (m^{-3} s), V_{d} is the deposition velocity (m s^{-1}), and τ is the effective mean residence time on grass (~5 days). T_{r} as used in Table 3-1 is the product of A, the area of grass grazed per day (m^2), f_R, the fraction of radioiodine ingested by the cow per unit area grazed and, f_T, the fractional transfer to milk per unit volume. Estimates of typical values for V_{d}, τ, f_T are available from special studies such as those summarized by Eisenbud (1973) and UNSCEAR (1972). Measurements of mean air concentration (χ) or of mean total deposition ($\chi V_d \tau$) = G (m^{-2}) on grass can thereby be related to milk concentration and dose. By combining Eqs. (3-1) and (3-2), the relative uncertainties of dose measurement by the three types of measurements, milk concentration (C), total deposition (G), and air concentration (χ) can be illustrated, i.e.,

$$\dot{D} = \mathrm{k}I\, \mathrm{f_T f_R}\, A\, \chi\, V_d\, \tau \tag{3-3}$$

It should be noted that measurements of χ may be useful in evaluating the alternate dose pathway, inhalation. This would be important when fresh milk is not consumed.

Although measurements of χ in air or G on grazing lands can be used to calculate milk concentration and the consequent absorbed dose rate, measurements of the concentration of radionuclides in milk lead to more accurate estimates of the absorbed dose rate as can be seen from Eq. 3-3. If χ is measured, the absorbed dose rate is the product of the 7 factors multiplied by χ. To avoid underestimating the dose due to uncertainties in measured factors, conservative values can be used.

The proper approach is described by Lengemann (1966) for a single contaminating event. An integrated form of Eq. 3-1 relates total expected human ^{131}I intake to the measured concentration of ^{131}I in milk at any time within about 15 days. This ratio is called F and is used in the integrated form of Eq. 3-1 to give the total expected absorbed dose.

$$D = \mathrm{k}I\, C\, F \tag{3-4}$$

where C is determined at the time after the event for which the F value is appropriate. D is then a type of absorbed dose commitment from a single contaminating event, which can be estimated from a single determination of ^{131}I concentration in milk. Lengemann also describes how to modify the dose estimate in the event that cows are removed from access to contaminated pasture.

Similar observations could be made for other dose pathways; however, the present discussion is sufficient to suggest guidance for environmental measurements:

(a) The goal of environmental measurements directed toward assessing the dose from internally deposited radionuclides is to quantify their human intake rate and resulting organ concentration;

(b) Most measurements will be directed toward defining radionuclide concentrations in one or more environmental media;

(c) In the absence of comprehensive dietary measurements, the experimental data must be supplemented by a computational model of radionuclide transport in the environment to man;
(d) The identification of critical dose pathways is a prerequisite to the design of a rational measurement program and this identification requires a detailed specification of the radionuclide source;
(e) The choice of measurements to be made depends on the degree of accuracy desired in dose assessment, on practical limitations (e.g., sensitivity, discrimination, cost) of available techniques, and on the importance attached to the characterization of actual environmental radionuclide distributions;
(f) In the absence of direct measurements of man himself, the media consumed most directly by man should be measured to obtain optimal accuracy.

3.3.2 *Dose From Externally-Incident Radiation*

The inference of dose to man from measurements of environmental radiation fields or radioactivity distributions involves the interpretation of measured quantities in terms of the radiation field and the application of appropriate dose models which combine the physical parameters of the radiation field and the appropriate whole body or organ geometry.

Interpretation of detector response involves discrimination among the various components of the radiation field, described in Section 2.3 and the following sections, e.g., cosmic-ray charged particles and photons from various sources and proper application of detector calibration in terms of the field. The characteristics of a particular environmental field, the composition, intensity, energy distribution, and angular distribution, cannot usually be measured, but they can be inferred from other information. For example, measurements of the concentrations of natural radionuclides in near-surface soil at an outdoor location (see Figure 2-4), coupled with the reasonable assumption of a homogeneous distribution with depth, is usually adequate to determine the natural gamma radiation absorbed dose rate in air above the soil (e.g., see Section 2.3.3 and Gustafson and Brar, 1964).

The inference of human organ dose from a knowledge of the environmental radiation field requires the application of the appropriate conversion factors (e.g., ratio of absorbed dose in some organ to that in free air), such as those determined from gamma radiation by

O'Brien *et al.* (1958); Spiers and Overton (1962); Jones (1966); and Bennett (1970).

A more direct approach to dose assessment would be to use a detector whose response is directly proportional to a particular human organ dose. An example of this might be a measurement inside a human phantom. This approach is probably impractical and additional measurements would be needed for determining the parameters of the radiation field.

Realistic long-term estimates of absorbed dose to individuals or to averages of populations should take into account their movement and living habits. For example, determinations of natural gamma- plus cosmic-radiation absorbed dose rates at outdoor locations do not completely define the radiation environment of most individuals, much of whose time is spent indoors. The outdoor radiation field is modified by the degree of shielding provided by buildings and by the additional field generated by the radionuclides in their materials (NCRP, 1975). The effects of these modifications vary considerably from structure to structure and the outdoor field can vary substantially with location and time. The design of a measurement program for the determination of population absorbed dose should account for these factors.

There are two feasible approaches to population absorbed dose assessment from external radiation. *In situ* measurements of the radiation field can be made at "representative" indoor and outdoor locations, and the results weighed according to typical or average occupancy times. Alternatively, small sensitive dosimeters can be worn by a "representative" sample of the population for sufficiently long periods of time to average the occupancy times in various radiation environments. This second method cannot be easily applied to the assessment of the absorbed dose contributions from particular components of the total radiation environment, e.g., from man-made radionuclides.

3.4 Surveillance Around Nuclear Facilities

3.4.1 *Objectives*

Routine environmental radiation monitoring programs should be designed specifically for each facility. The complexity of a program depends on the quantities, chemical forms of radionuclides, identities, and physical forms of radionuclides that may be released and on the characteristics of the environs of the facility. However, "routine" measurements must be adequate to determine compliance with environmental standards, the parameters needed for subsequent dose assessment, and trends of environmental radioactivity and radiation levels.

The objectives of routine surveillance programs include:

(a) Providing information useful in assessing the adequacy of protection of the public;
(b) Meeting requirements of regulatory agencies;
(c) Verifying radionuclide containment and plant waste management practices;
(d) Meeting legal liability obligations; and
(e) Providing public assurance and acceptance.

There is always a minimum detectable concentration or dose associated with a sample collection and analysis or an *in situ* measurement. Accordingly, measurement methods should be selected or developed so that this detection limit corresponds to a dose to the critical group in the critical path that is below some predetermined acceptable value recommended by a scientific commission or established by a regulatory agency.

3.4.2 *Development of Environmental Surveillance Programs*

General criteria for the design of environmental surveillance programs have been described by the ICRP (1965a; WHO, 1968). Those criteria directly related to the choice of measurement techniques and procedures have been indicated in the preceding sections.

The following principles for the development of a practical surveillance program for a nuclear facility involve steps that are reasonably independent of local circumstances:

(a) Evaluate the facility as a source of direct radiation and radionuclides, especially the composition, concentrations, release rates, points of release, and physical and chemical forms of the nuclides.
(b) Identify the pathways leading to exposure to man from source data by the use of analytical models for the possible pathways, and by use of experience gained at other sites and preoperational data on local meteorology, hydrology, and population distribution and diet.
(c) Select the significant pathways that may be critical in terms of their contributions to exposure and determine the critical population groups.
(d) Determine the consequent measurement requirements to pro-

vide data for dose assessment for both normal and abnormal conditions.

(e) Determine other measurement requirements.

(f) Allow for flexibility in the design as operational experience is accumulated and indicates that other types of measurements or measurement frequencies may be desirable as the dose pathways and the other program requirements become better defined.

3.5 Dose to Population Groups

In their considerations of the limitation of dose to the general public from ionizing radiation, the FRC (1960), the NCRP (1971) and ICRP (1965b) distinguish between dose to individuals, to critical groups, and to the general population. The ICRP (1965a) defines the critical group as that "whose exposure is homogeneous and typical of that of the most highly exposed individuals in the exposed population." Because of the difficulty in determining maximum doses to individuals in particular situations, one should select appropriate critical groups and apply the dose limit recommended for individual members of the public to a suitable sample of the group. The NCRP (1971) does not specifically make this distinction and recommends dose limits only for individuals and for the population of the United States as a whole. The Nuclear Regulatory Commission referred to dose to individuals and to "sizeable population groups" in connection with the development of design objectives for light-water power reactors (USNRC, 1975; USAEC, 1973a). It should be noted that the NCRP and ICRP dose limits refer to all sources of radiation except natural background and medical radiation, while the NRC limits represent an attempt to quantify "as low as practicable" values for that specific industrial activity. The NCRP and ICRP dose limits refer to the annual dose to a member of a population group, a maximum dose in the case of individuals and mean dose for critical groups and the general population. Another measure of the possible impact of a radiation dose to a population is the total dose delivered to all members of that population. This parameter is equal to the product of the mean or individual dose and the population size and its use might require a modification of the nature and scope of monitoring programs.

4. *In Situ* Radiation Measurements

4.1 Introduction

In principle any measurement of environmental radioactivity involves the measurement of a sample. When an *in situ* method is employed, the sample retains the character of the environment.

In situ measurements are valuable for the rapid assessment of radiation exposure, identification of radionuclides, and detection of trends in environmental radioactivity due to man's activities. Measurements are made to evaluate natural sources (Adams and Lowder, 1964; Neher, 1958); the extent of mineral deposits (Adams and Gasparini, 1970); and perturbations due to soil moisture, snow cover and forests (Kogan *et al.*, 1969).

Environmental gamma-ray monitoring is performed near nuclear facilities to assure that dose contributions comply with acceptable values (ICRP, 1965a; 1965b; NCRP, 1971). The measurement methods described provide guidance for designing a monitoring program.

Gamma-ray measurements provide data for identifying and locating sources in the environment and estimating doses. The total gamma-ray ionization rate is used to estimate the dose rate to the population. *In situ* spectrometry is used to identify radionuclides in the environment and estimate their concentrations.

Detectors that accumulate radiation-induced effects over specified time periods are convenient because they are small, field work is minimal, and measurements are performed under controlled laboratory conditions. The determination of radiation exposure or dose in this fashion usually precludes evaluation of small or short-term changes, although statistical comparisons of many measurements may indicate significant temporal or spatial changes (Shapiro *et al.*, 1968).

A measurement program aims at the identification of the radiations and the selection of appropriate instruments. Nuclear facilities monitoring programs were originally carried out to determine if additional radiation dose rates or radioactivity concentrations were within accepted values and to provide a measurement capability in the event of large radioactivity releases into the environment. Recommended maximum values corresponded to an annual whole body dose equivalent of 0.5 rem above natural background and it was recognized that actual levels should be kept to some lower value (FRC, 1960; ICRP, 1965b). The NCRP indicated that "it may, at some time, become reasonable or necessary to stipulate all permissible dose limits in terms of total dose, rather than to continue the practice of permitting some dose in addition to natural background (NCRP, 1971)". Such considerations imply that knowledge of natural background variations and the small contributions from manmade radionuclides is needed. If so, there may be a requirement for improvements in the reliability and sensitivity of measurement methods. Instabilities in instrument response should not be as large as either the background variations or routine contributions by the facility being monitored.

If the purpose of an environmental measurement program is to characterize the total gamma radiation field within an overall uncertainty of ±20 percent, there are correspondingly stringent requirements on the precision of instrument response and accuracy of calibration. Some instruments are capable of measuring the average exposure rate with a smaller uncertainty. Thus low-level environmental monitoring and detection of long-term variations and trends require that minimal uncertainties should be maintained over long periods (Beck *et al.*, 1972a).

Accurate measurements of either the total environmental dose rate or the contributions from particular manmade sources require that account be made of the contributions from natural terrestrial radiation, long-lived manmade radionuclides and cosmic radiation. Acceptable instruments and problems in their use are described and appropriate literature is cited in the following sections.

4.2 Ionization Chambers

4.2.1 *Historical Development*

The initial nuclear facilities environmental radiation measurements were made with cylindrical air-filled chambers (Parker, 1946; Kuper and Chase, 1950). The chambers were charged and exposed, and the consequent discharge was correlated with exposure, based on calibration with a standard source. Wall thicknesses were selected to inhibit the response to beta rays, but thin windows were sometimes employed to obtain approximate information on relative gamma- and beta-ray responses. An automatic recording system was designed by Kuper and Chase (1950).

Early environmental radiation measurements with ionization chambers have been described by Hultqvist (1956), Libby (1955), and Solon *et al.* (1959). Extraneous currents produced in low pressure, air-filled chambers by internal alpha radioactivity and temperature dependence caused by variations in the air mass led to the use of high pressure ionization chambers by Hultqvist (1956) and Spiers *et al.*, (1964). Previously Millikan (1932), his successors (Neher, 1958) and others made cosmic-ray measurements with similar chambers. Due to the inability to obtain saturation in high pressure air chambers, argon or nitrogen was used; the required thick walls introduce an effect for which correction must be made.

Rose and Shonka (1968) developed a 16-liter muscle tissue equivalent chamber used in conjunction with a vibrating quartz fiber electrometer capable of measuring typical environmental exposure rates (Kastner *et al.*, 1964). Shamos *et al.* (1964) and Shamos and Liboff (1968) showed, however, that the effect of alpha activity from chamber walls, which depends on surface area, may be significant in these large volume, low pressure chambers. Shamos *et al.* (1964) developed a low pressure chamber containing the electronegative gas, Freon-12 (CCl_2F_2), leading to reduced noise fluctuations and alpha-emitting radioactivity without significant reduction in sensitivity. Both of these chambers have fairly thin walls and the consequent response to environmental beta particles should be distinguished from that from gamma-rays, possibly by increasing the chamber wall thickness.

4.2.2 *Gamma-Ray Response*

Ionization chambers are used to measure the exposure rate from gamma emitters in the ground at a nominal height (one meter), which information is then used to estimate absorbed dose rate to man as discussed in Section 3. For accurate measurements of exposure rate, it is important to determine measurement uncertainties carefully by performing both laboratory and field calibration.

Guidance to the selection of chambers and their calibration is obtained by considering the response to monoenergetic gamma-rays. The gamma-ray response, k_γ, relates the instrument output to exposure rate or absorbed dose rate in free air under a condition of electronic equilibrium, that exists in free air to a close approximation in environmental radiation fields. Neglecting any wall effect, the gamma-ray response, k_γ, is estimated from μ_{tr}/ρ, the mass energy transfer coefficient for photons (NAS-NRC, 1964); ρ, the gas density (Weast, 1971); and $\bar{W}$, the mean energy required to form an ion pair (ICRU, 1964). In Table 4-1, the k_γ for a number of chambers shows the range of sensitivities to a typical gamma-ray field of the type indicated in Table 2-10. The μ_{tr}/ρ values account for the energy transferred to electrons per unit path length, but in a practical chamber energy will be carried away from the sensitive volume by energetic electrons and bremsstrahlung. In a uniform, large medium, this fraction is compensated by energy introduced from the surroundings.

For an argon-filled chamber,

$$k_\gamma = C_1 \frac{(\mu_{tr}/\rho)_{Ar}}{(\mu_{tr}/\rho)_{air}} (PV)_{STP} \tag{4-1}$$

where C_1 is 0.164 fA/μR h^{-1} per liter-atmosphere[1], neglecting any wall effect, and P is in atmospheres at 0°C and V is the effective volume of the chamber in liters. For a similar nitrogen-filled chamber, like that of Spiers *et al.*, (1964), C_1 is 0.0914 fA/μR h^{-1} per liter-atmosphere. The interpretation of measurements in terms of exposure rate or absorbed dose rate in free air depends on general considerations of cavity-chamber theory developed by Gray (1936); Spencer and Attix (1955); and Burlin (1966), and comprehensively discussed in NCRP (1961) and Burlin (1968). Although most frequently applied to ionization chambers, such considerations also apply to other detectors, e.g., solid state dosimeters (Burlin and Chan, 1967).

Eq. 4-1 must be modified to account for attenuation of the incident photon flux density by the chamber wall and production of scattered photons, as well as for the difference in energy transferred to secondary electrons between the wall and gas material per unit path (g cm^{-2}). Because the latter effect depends on atomic number, it is higher, for example, in steel than in argon and an excess of ionization is produced in the gas per unit weight. De Campo *et al.* (1972) have represented these two effects as follows:

$$k_\gamma = 0.164\ F(E,X,PV)\ PV_{STP} \tag{4-2}$$

where $F(E,X,PV) = B_e B_\gamma \frac{(\mu_{tr}/\rho)_{Ar}}{(\mu_{tr}/\rho)_{air}} e^{-(\mu/\rho)X}$, B_e is the multiplicative buildup factor from excess electron production in the wall, B_γ is the photon ionization buildup factor, i.e., the ratio of ionization from all photons to that from externally-incident photons, μ/ρ is the mass attenuation coefficient for the externally-incident photons in the

[1] The quantity pressure (P) in atmospheres is retained here, because of its extensive use in the prior literature (see references in Table 4-1 and U.S. Standard Atmosphere, 1962). 1 atmosphere = 1.01×10^5 Pa (Pascal).

wall, and X is the effective wall thickness in g cm^{-2}. The dimensionless values for F(E,X,PV) in Table 4-2 from DeCampo *et al.* (1972) for steel-wall, argon ionization chambers are useful for selecting a chamber with a response appropriate for a particular energy spectrum. The effective wall thickness in the table is about 1.2 times the perpendicular thickness in spherical chambers. Similar F values should be developed for other high pressure chambers.

As Table 4-2 indicates, the response of an argon-filled chamber with a 2.5 g cm^{-2} iron wall is reasonably constant above about 0.15 MeV due to the compensating effects of wall attenuation and argon energy absorption (De Campo *et al.*, 1972). The response of a thin wall argon chamber is less "flat" and unsuitable for environmental meas-

TABLE 4-1—*Ionization chamber responses*

Gas	Pressure	Volume	Wall thickness	Gamma-ray response	Reference[a]
	atm	l	g cm^{-2}	A/μR h^{-1}	
Air	1	20	1.08	2.1×10^{-15}	1
Nitrogen	45	5.5	2.54	3.5×10^{-14}	2
Argon	59	5.6	2.74	4.5×10^{-14}	3
Argon	20	8	2.37	2.4×10^{-14}	4
Muscle Equivalent	1	16	0.64	1.6×10^{-15}	5
Freon	1	16	0.52	5.8×10^{-15}	6

[a] References: 1—Solon *et al.*, (1959); 2—Spiers *et al.* (1964); 3—Beck *et al.* (1966); 4—De Campo *et al.* (1972); 5—Rose and Shonka (1968); 6—Shamos and Liboff (1966).

TABLE 4-2—*Calculated F(E,X,PV) for high pressure argon, steel ionization chambers*

Energy	F(E,X,PV) for various effective wall thicknesses (g cm^{-2})[a, b]							
	0.5	1.0	1.5	2.0	2.5	3.0	3.5	4.0
keV								
30	0.317	0.007	—	—	—	—	—	—
50	5.08	2.08	0.854	0.340	0.138	0.056	0.028	0.014
60	5.66	3.32	1.92	1.11	0.637	0.370	0.210	0.113
80	4.33	3.44	2.70	2.12	1.66	1.29	0.986	0.766
100	2.89	2.59	2.28	1.99	1.71	1.48	1.27	1.09
150	1.51	1.48	1.42	1.36	1.30	1.23	1.15	1.07
300	0.979	0.986	0.988	0.973	0.967	0.967	0.934	0.912
500	0.925	0.940	0.941	0.932	0.934	0.932	0.914	0.861
1000	0.913	0.903	0.901	0.890	0.890	0.877	0.865	0.862
2000	0.918	0.904	0.897	0.887	0.884	0.881	0.870	0.861
3000	0.958	0.949	0.942	0.933	0.926	0.926	0.919	0.910
4000	1.00	0.996	0.980	0.973	0.966	0.969	0.961	0.955
6000	1.09	1.09	1.07	1.06	1.06	1.05	1.03	1.03
8000	1.16	1.14	1.15	1.12	1.12	1.10	1.10	1.08

[a] From De Campo *et al.* (1972).

[b] The tabulated values are for $B_e = 1$, i.e., $PV > 200$ liter—atmospheres. At $PV \approx$ 50 liter-atmospheres, for example, B_e can be as large as 1.1 near 50 keV and for high energies, ~ 5000 keV.

urements because of the greatly enhanced low-energy response. De-Campo *et al.* (1972) have compared the response of different high-pressure, argon-steel chambers to primary photons from a ^{226}Ra calibration source and a typical environmental gamma radiation field. The comparisons showed discrepancies well within the desired ±20 percent (see Section 4.1), even for thin wall chambers, and about 1–10% for thick walls (De Campo *et al.*, 1972). The radium spectrum obtained from Table 2-2 and the environmental spectrum from Beck (1972b) are shown in Figure 4-1 and indicate the need for careful calibration if the detector response is not uniform with energy (see Section 4.2.4 and 4.3.1). The ratio of the response to the calibration source and that to expected environmental sources should be essentially unity. The function, F(E,X,PV) can also be inferred for other common calibration sources, e.g. ^{137}Cs and ^{60}Co, from the data in Table 4-2.

4.2.3 *Cosmic-Ray Response*

Cosmic-ray secondaries produce a large part of the free air ionization that one measures in the lower atmosphere (see Section 2.3.2.3).

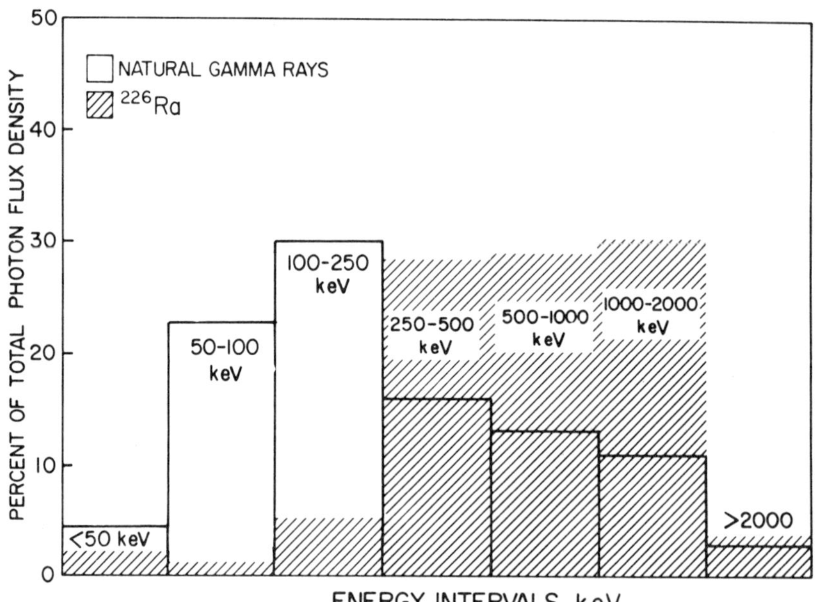

Fig. 4-1. Comparison of ^{226}Ra calibration source energy spectrum with that from environmental natural gamma rays (latter from Beck, 1972b).

The determination of the cosmic-ray response is necessary for the determination of the terrestrial gamma-ray component. The free air ionization rate from cosmic radiation[2] (see Section 2.3.2.3) in the lower atmosphere depends mainly on the mass of air above the detector, i.e., the atmospheric pressure, and varies by about a factor or two in the populated altitudes (see Figure 2-2).

The cosmic-ray response k_c is expressed much like the gamma ray response in Eq. 4-1, except charged-particle collison stopping-power values replace the photon energy transfer coefficients. The response is expressed in amperes per unit ionization rate in free air at STP. The overall stopping power ratios for common gases given in Table 4-3 were calculated from reasonable approximations for the muon and electron spectra (Lowder, 1969; Grotch, 1962) and published stopping powers (NAS-NRC, 1964). For argon, applying the value from Table 4-3,

$$k_c = 0.85\ C_2\ (PV) \tag{4-3}$$

where C_2 is 0.284 fA per unit ion pair $cm^{-3}\ s^{-1}$ in free air per liter-atmosphere at STP, wall effects being neglected. Chambers such as the Rose and Shonka (1968) tissue equivalent and the Shamos and Liboff (1968) freon-12 systems have relatively thin walls and wall effects are small and probably negligible.

The muon component produces most of the ionization at sea level and is attenuated very little by chamber walls less thick than a few g cm^{-2}. A small proportion of the electron-photon component is sufficiently energetic to produce electromagnetic showers in air. This shower production is enhanced in higher-Z materials, such as iron, and results in an increased electron flux density (Beck, 1971). Experiments by Hess and Manning (1956), Clay (1939), and Hopfield (1933) indicate about a 10 percent ionization enhancement for chambers with thick steel walls. Experiments with 17.8 cm and 25.4 cm diameter steel-argon spheres show the wall effect to be less than 10 percent for steel wall thicknesses up to 2.7 g cm^{-2} (De Campo *et al.*, 1972) and are consistent with calculations (Beck, 1971).

The appropriate stopping power ratios from Table 4-3 may be used to determine k_c in Eq. 4-3, neglecting any wall effect. This effect is then estimated from calculations (Beck, 1971).

[2] This quantity can be converted to absorbed dose rate in free air by the relation 1 ion pair $cm^{-3}\ s^{-1} = 1.50\ \mu rad\ h^{-1}$, since secondary particle equilibrium approximately pertains.

TABLE 4-3—*Mean collision stopping powers of common ionization chamber materials for sea-level cosmic-ray charged particles*

Material	Electron stopping power	Muon stopping power	Gas to air ratio[a]
	MeV cm² g⁻¹	MeV cm² g⁻¹	
Air	2.21	2.25	1.0
Argon	1.79	1.92	0.85
Nitrogen	2.12	—	0.96[b]
Tissue equivalent gas (muscle)	2.34	—	1.05[b]
Freon ($CF_2\ Cl_2$)[c]	1.99	2.11	0.94

[a] Assuming 70 percent of cosmic-ray ionization at sea level is due to muons and 30 percent to electrons (see Section 2.3.2).

[b] Muon stopping power ratio assumed equal to that of electrons.

[c] Approximate values based on Brandt (1960) data from 0–10 MeV.

4.2.4 *Calibration for Field Measurements*

Ionization chamber calibrations for low level environmental measurements require more care than is usually necessary for radiation protection instruments. The response as a function of energy should be reasonably well known, and preferably constant with energy or nearly so.

The total chamber current is

$$i_t = k_\gamma \dot{F}_\gamma + k_c \dot{F}_c + i' \qquad (4\text{-}4)$$

where k_γ and k_c are obtained as described in the previous two sections (in appropriate units), $\dot{F}_\gamma$ and $\dot{F}_c$ are common field quantities[3] for gamma and cosmic rays, respectively, usually absorbed dose rate in free air or air ionization rate, and i' is the current contribution from radioactivity, usually alpha emitters, inside the chamber, plus leakage and stress currents. i' should be negligible in well-designed and maintained systems.

The energy response of the chamber should be determined from the data in Table 4-2 for steel-argon systems, or as adapted for other fillings by recourse to Eq. 4-2 using published $\bar{W}$ and μ_{tr}/ρ values. The low-pressure, tissue-equivalent chamber has a nearly "flat" response

[3] It is sometimes desirable to express the quantitative properties of cosmic rays and terrestrial gamma rays in common units. Since $\dot{F}_c$ is rarely measured, but is obtained from the literature as ionization rate or absorbed dose rate in free air, k_c is conveniently expressed in such units. $\dot{F}_\gamma$ and k_γ most frequently involve the quantity, exposure rate, one that is inapplicable to cosmic-ray charged particles. Because secondary charged particle equilibrium usually exists in environmental gamma-ray fields, the conversion of exposure rate to quantities appropriate for the cosmic-ray field can be made by 1 μR h^{-1} = 0.869 μrad h^{-1} = 0.577 ion pairs cm^{-3} s^{-1} at standard temperature and pressure, all in free air.

and the tedious experimental determination of the energy response may be unnecessary.

Account should be taken of the effect of scattering by material in the calibration area and the source should be placed far enough from the center of the chamber to minimize the effect of large chamber dimensions. The total response is attributable to primary and secondary photons from the source and from laboratory background, plus any contribution from internal radioactivity. Laboratory background includes the room scatter component, which is geometry dependent and often significant. Correction for this is made by carefully performing a shadow shield experiment to determine the ionization due to primaries alone. As it is not possible to shield many chambers exactly with the usual lead bricks, a slight amount of overshielding is preferred. Undershielding can lead to an overestimate of the ionization attributed to scattered photons. The position and size of the shield or the source-to-detector distance should be varied to assure that the correct procedure is being followed in terms of yielding values of response per unit primary flux density.

If the chamber gamma-ray response is known, one need only measure ionization produced by a known source, such as a sealed 1-mg ^{226}Ra source that produces 0.825 mR h^{-1} at one meter (ICRU, 1971b), and then infer the required correction for a different energy spectrum. For the high-pressure argon chambers, only a few percent correction is needed to convert a ^{226}Ra calibration to that for the environmental spectrum shown in Figure 4-1 (De Campo *et al.*, 1972).

Consideration should be given to any contributions from internal radioactivity. Stainless steel emits alpha particles sufficient to produce approximately 10^{-15} amperes of excess current in a 17.8-cm diameter spherical chamber and about twice this amount in a 25.4-cm chamber (Price, 1958). This ionization current remains constant above a gas pressure sufficient to stop alphas, while that due to gamma rays increases linearly with pressure, so the correction for internal alpha ionization can usually be neglected in high pressure chambers (De Campo *et al.*, 1972). Nevertheless, newly procured chambers should be checked for unusual contamination from manufacturing processes. Various tests have been described by Shamos and Liboff (1966), Solon *et al.* (1959), and George (1970).

One should calibrate the entire instrument system, but in large monitoring programs separate calibration of major components permits them to be interchanged. Electrometers used with ionization chambers may function as voltmeters having known input resistances which should be much greater than any extraneous resistances. The ionization current may also be determined by measuring the time rate of voltage change across a known capacitor with the known resistance removed from the electrometer input.

The value of the input resistance used in the first method may change due to aging, temperature and humidity, and noise, and periodic tests are advisable. Spiers *et al.* (1964) and George (1970) and others have described their versions of the second method.

4.2.5 *Field Measurements*

A reasonably flat, undisturbed site, and assurance it will remain so, is recommended for repeated field measurements. Calculations by Minato (1971) and Beck (1972b) showed that most of the flux density and the exposure rate one meter above ground containing uniform concentrations of ^{40}K and nuclides of the ^{238}U and ^{232}Th series are produced by the sources within a 5-meter radius. Sensitive measurements, usually with survey instruments, should be made to verify that the site is uniform and representative.

Attention should be paid to the expected local energy spectrum in special monitoring experiments, such as that due to the gaseous effluent from a nuclear power station (McLaughlin and Beck, 1973) and/or the direct radiation from ^{16}N ($\sim$6 MeV) in the steam turbine of direct cycle plants (Lowder *et al.*, 1973; Holt and Gibson, 1964). Unusual energy spectra, such as these, often require supporting spectrometric measurements to aid data interpretation.

When dose to man is the quantity of interest, measurement sites should be near populated areas, but away from local non-representative radioactive sources, such as rock outcroppings. The determination of dose to man from measurements in uninhabited locations and the inference of long-term dose from instantaneous or short-term measurements should be avoided, since they may lead to substantial errors.

Terrestrial gamma-ray levels are affected by many factors (see Section 2.3.3). For monitoring around a nuclear facility, long-term measurements at a relatively few sites are to be preferred over infrequent instantaneous measurements, because extrapolation of the latter to annual doses will very likely be erroneous. This problem has indicated the need for reliable and sensitive continuous monitors like the systems described by Beck *et al.* (1972a), Jones (1974) and Chester *et al.* (1972).

To account for the cosmic-ray contribution, one can measure atmospheric pressure and subtract the instrument response corresponding to the absorbed dose rate from Figure 2-2 to obtain the gamma-ray contribution.

Another factor in planning field measurements is electrometer performance. The electrometers used by Spiers *et al.* (1964) and by Shonka (1962) relied essentially on charge-balancing circuits where the voltage change with time determines the exposure rate. The terrestrial and cosmic ray measurements of Shamos *et al.* (1964); Beck *et al.* (1966); and Solon *et al.* (1959) were made with commercial vibrating reed electrometers, powered by batteries. These electrometers are essentially laboratory instruments, so they are not suited to long term field monitoring.

New solid state electrometers having high current sensitivity like the device described by McCaslin (1964) show much promise for field work. Negro *et al.* (1974) have described a MOSFET (metal oxide semiconductor field effect transistor) electrometer. High impedance between input and ground insures a low leakage current (about one fA) and the temperature dependence from −23°C to 54°C corresponds to a current shift of only 5 fA. The electrometer voltage output, which corresponds to the exposure rate, is recorded on magnetic tape cassettes and permits automated data reduction and long-distance transmission of results (Cassidy *et al.* 1974).

The responses described in Section 4.2.2 pertain to bare chambers, but field systems should be packaged to provide protection against dirt, moisture and damage. Such packaging can affect the energy response and, in addition, the location of the electrometer and batteries can affect the angular response.

The overall accuracy of free air exposure rate measurements can be very high. Rose and Shonka (1968) demonstrated that the calculated and measured response of tissue- and air-equivalent chambers to a ^{226}Ra source agreed within ± 0.5 percent in the laboratory; the overall uncertainty is somewhat larger for field measurements. De Campo *et al.* (1972) indicated the overall uncertainty in pressurized-argon chamber measurements to be about ± 5 percent.

4.3 Portable Scintillation and G.M. Instruments

4.3.1 *Problems*

Instruments employing scintillation or G.M. counters are widely used for measuring radiation from naturally-occurring sources (e.g., Wollenberg and Smith, 1964; Levin and Stoms, 1969; Aten *et al.*, 1961). Due to the energy independence of proportional counters above 100 keV, they show promise and, as indicated by Nakashima and

Kawano (1972), measurements agree well with those from ionization chambers.

An ideal survey or monitoring instrument should:

(a) be sufficiently sensitive to allow the measurement of environmental gamma rays down to at least 1 μR h^{-1};

(b) have a sufficiently short time constant to permit the making of many measurements, but sufficiently long to minimize rapid fluctuations,

(c) have a uniform and nearly isotropic response over the gamma-ray energy range of interest,

(d) be insensitive to the rigors of field use and, particularly, to temperature changes, and

(e) be small and lightweight, whether it is used for hand-held surveys or for stationary monitoring.

Practical instruments do not exhibit all these characteristics.

An energy independent response can be achieved by using detector shields (Aten *et al.*, 1961; Emery, 1966). Jones (1974) describes a monitoring instrument employing an energy-compensated G.M. counter with a readout displayed on a logarithmic scale and on a register. The instrument operates in the range of 5 μrad h^{-1} to 5 mrad h^{-1} and the minimum detectable absorbed dose is 1 μrad. The sensitivity of the system indicates it is suitable for monitoring routine releases close to nuclear facilities and for abnormal releases.

Sensitive scintillation detector instruments have received the greatest use as survey instruments (Wollenberg and Smith, 1964; Wensel, 1964; Levin and Stoms, 1969). The counting rate per unit exposure rate produced in these instruments is energy dependent, this dependence being determined by the energy absorption coefficients of the elements that comprise the scintillator.

Organic scintillators have effective atomic numbers nearly equal to that of air, although at low photon energies they exhibit a decrease in response, which depends on their thickness (Bland and Schoenaich-Peters, 1972; Kolb and Lauterbach, 1972). Inorganic scintillation detectors, usually NaI(Tl), show enhanced low energy responses because of their greater effective atomic numbers. Johnson (1972) and Beck *et al.* (1972b) showed that the responses of the standard 5.1 × 5.1-cm and 10.2 × 10.2-cm NaI(Tl) detectors are markedly energy dependent. Two interpretation problems result from departures from uniform response with energy and angle and from the greater sensitivity to low energy photons, since the tacit assumption is made that the total counting rate above some discriminator threshold level is proportional to exposure rate. By setting the discriminator sufficiently high, one assures that gain shift or zero shift will have a negligible effect on the response. This is important because the environmental low energy photon flux density is relatively large and variable (Beck, 1972b). In addition, the determination of exposure rates based on calibrations with radioactive sources such as ^{137}Cs, ^{60}Co or ^{226}Ra can be seriously in error. Figure 4-1 shows that the spectrum from encapsulated ^{226}Ra in equilibrium with its daughters, as inferred from Martin and Blichert-Toft (1970), differs significantly from a typical natural, environmental spectrum one meter above ground as obtained from Beck (1972b). An instrument calibrated with the "harder" ^{226}Ra spectrum and then used for determining natural environmental exposure rates may yield erroneously high results.

Kolb and Lauterbach (1974) employed an anthracene detector, having a photon response proportional to exposure rate above about 100 keV. The low anthracene sensitivity to low energy gamma rays was compensated by coating the scintillator with ZnS(Ag) to make the overall response air equivalent above about 20 keV.

Energy independent plastic scintillators were described by Chester *et al.* (1972) and Bland and Schoenaich-Peters (1972), who employed tin-loaded and lead-loaded plastic scintillators, respectively. Development in this area is desirable for the expected need for environmental monitoring. The tin-loaded scintillator system covers an exposure rate range of 1 μR h^{-1} to 10 mR h^{-1}, making it useful for monitoring levels near background as well as those from sizeable radionuclide releases.

4.3.2 *Calibration for Field Measurements*

Survey instruments should be calibrated as suggested in Section 4.5.2 for gamma-ray spectrometers. Account should be taken of the detector energy and angular response and of departures of the photon flux density energy and angular distributions in the environmental radiation field from those of calibration sources.

Calibration factors are determined by exposure to known sources in the laboratory and the resulting survey instrument response compared to a measurement with a well-calibrated ionization chamber (Section 4.2.4). The measurement should be corrected to account for the fact that photons from the point source, unlike those in the environment, are generally incident along one direction. One may also calibrate with simultaneous *in situ* spectrometer measurements for different environmental radiation fields (Beck *et al.*, 1972b). Similarly, one can calibrate the G. M. or scintillation counter in the field with a suitable ionization chamber, provided care is taken to correct for the response of each to cosmic radiation.

Levin and Stoms (1969) employed basically the latter technique during a survey of background dose rate in the mid-United States, by comparing their survey meter with a tissue equivalent ionization chamber, which in turn had been calibrated with a ^{226}Ra source.

One should avoid assuming incorrectly that the gamma-ray and cosmic-ray responses of a particular detector are equivalent. An experimental evaluation of the cosmic-ray response can be made at various altitudes by measurements at locations with very low gamma background, such as on large bodies of water. Alternatively, if the theoretical response to cosmic rays per unit absorbed dose rate in free air can be estimated, the response as a function of pressure-altitude can be determined from the data given in Figure 2-2.

4.4 Thermoluminescence Dosimetry

4.4.1 *Advantages and Problems*

Small integrating dosimeters are convenient because many measurements can be made simultaneously and relatively economically with the resulting total exposures determined in the laboratory.

The curiosity of Daniels *et al.* (1953) marked the beginning of practical thermoluminescence dosimetry (TLD) with lithium fluoride and the work by Schulman *et al.* (1951), utilizing synthetic manganese activated calcium fluoride (CaF_2:Mn), led to a substantial and rapidly expanding literature on thermoluminescence dosimetry (e.g., see Attix, 1967; Auxier *et al.*, 1968; and Mejdahl, 1971; and the excellent texts by Cameron *et al.*, 1968; and Becker, 1973).

Photographic film has been employed for nuclear facilities environmental radiation monitoring, but film is not adequate for low-level measurements because of its relatively low sensitivity and susceptibility to damage by heat, moisture, and light (Parker, 1946; Fitzsimmons *et al.*, 1972). Observations of submicroscopic tracks in thin plastic films and mica produced by heavy charged particles (hadrons and fission fragments) provide a valuable detection mechanism for radiation protection and geophysical investigations (Fleischer *et al.*, 1965). The measurement of radon and radon-daughter alpha tracks in plastic films placed in the ground and inside buildings appears promising for geophysical and population radiation exposure surveys (Gingrich and Lovett, 1972). Measurements of environmental gamma rays and leptons with these films do not appear feasible. The primary use of track etch methods in environmental radiation studies has been in connection with the laboratory analysis of heavy elements.

Thermoluminescence has been widely exploited for a variety of purposes (Daniels *et al.*, 1953 and Schulman *et al.*, 1951). However, the use of TLDs for environmental radiation dosimetry requires special dosimeter selection, and quality control and assurance (Burke and McLaughlin, 1974). Like any *in situ* instrument, TLDs should be little affected by environmental conditions. The beta-ray component of the environmental radiation field near the ground varies greatly with time, so the dosimeter should be shielded so as to detect only the penetrating component and to assure charged particle equilibrium (ICRU, 1964). It is advantageous for the effective atomic number of the dosimeter material to be close to that of air to minimize problems of interpretation. Because many TL phosphors are sensitive to visible and ultraviolet light, they should be protected during preparation, exposure, and readout.

Variations among dosimeters within a given production batch should be sufficiently small to avoid the requirement of a unique calibration for each dosimeter. Although the cost per dosimeter is small, tests for quality assurance will increase the number of dosimeters required. These tests should involve measurements with replicate and control detectors to assure that extraneous contributions during handling, readout, and shipment are properly determined.

The choice of gamma-ray sensitivities of phosphors is large (Binder *et al.*, 1968) although the dosimeter system response depends strongly on the exact readout instrument and procedures employed, including photomultiplier tube type, positioning and temperature stability, and phosphor composition. Low temperature thermoluminescence glow curve peaks are unstable, so that removal of such peaks by special annealing and reliance on higher temperature peaks is necessary for reliable low-level dosimetry. The infra-red emissions from heated components of the readout instrument and the dosimeter should be separable from the radiation-induced emission spectrum.

Two characteristics that affect accuracy and reliability are thermoluminescence fading with time and the contribution from any traces of radioactive material in the phosphor or packaging. These effects are usually negligible for short-term, relatively large exposures and timely readout, but not for protracted environmental exposures, followed by delays before readout. Fading depends on the phosphor itself and is reasonably constant among dosimeters from the same production batch. Self-irradiation due to trace radioactivity, however, may vary considerably among dosimeters. The evaluation of these opposite effects is difficult. They can be negligible for some materials when the dosimeter response is greater than the net contribution from these effects.

These effects can be described simply by

$$I_t = I'_t\, e^{-\lambda t} + \frac{\alpha \dot{X}}{\lambda}\,(1 - e^{-\lambda t}) \tag{4-5}$$

where

I_t = light intensity at time t,
I'_t = pre-exposure light intensity,
α = numerical conversion factor,
$\dot{X}$ = exposure rate due to the environment and any internal radioactivity, and
λ = thermoluminescence fading constant.

The evaluation of λ and the contribution by internal radioactivity to $\dot{X}$ require careful experimentation. Dosimeters should be exposed in a well-shielded location where the external radiation, due mostly to cosmic rays, is low and constant. The readouts, I_t, of subgroups of dosimeters removed from the shielded area at various times are fitted to Eq. 4-5, and λ and $\dot{X}$ determined. The external radiation contribution to $\dot{X}$ must, of course, be determined separately with a detector having negligible internal radioactivity, such as LiF, or with an ionization chamber.

The thermoluminescence contribution from trace radioactivity in many materials has not been found to be significant (Becker, 1973; Zimmerman *et al.*, 1966). The responses of calcium fluoride TLDs, due to natural radioactivity in the material and in any glass enclosure surrounding it, which corresponds to about 7–21 μR h^{-1} in CaF_2:Mn, may be significant (Aitken, 1968; Burke, 1972). Dosimeters having anomalously large contributions should be screened from batches.

Burke (1972) reported about 2.3 percent fading per month in the CaF_2:Mn dosimeters obtained for a simulated environmental exposure rate. Greater fading may be expected at the elevated temperatures of summertime (Becker, 1973).

Fading in irradiated LiF is generally considered to be very small at room temperature (Zimmerman *et al.*, 1966), although reported estimates vary. Becker (1973) estimated a fading rate of 3 percent for two months at 50°C and Shambon (1972) a maximum fading rate of about 3 percent per month at elevated summertime temperatures. Such results indicate that fading correction for LiF measurements may be neglected.

Taking account of the above problems, one can determine the average exposure rate from environmental gamma radiation usually within an accuracy of 5 percent.

4.4.2 *Typical Thermoluminescence Phosphors*

Other phosphors besides LiF and CaF_2 are suitable for low-level measurements. The higher atomic number phosphors are acceptable if the energy response is properly interpreted. Responses that have been calculated and measured by many investigators are summarized by Cameron *et al.* (1968) and Becker (1973). Materials like CaF_2 have greatly enhanced low-energy responses, though they are usually enclosed in filters to inhibit the beta and low-energy gamma-ray response. The responses of $Li_2B_4O_7$:Mn and BeO can be adjusted to be very close to that of air and tissue. Convenient solid forms have been developed and tested and are coming into use for low-level measurements. Some results with $Li_2B_4O_7$:Mn have been reported by Wilson and Cameron (1968) and with BeO by Becker (1973).

Binder *et al.* (1968) described the properties of compressed CaF_2: Dy which has been utilized in environmental monitoring programs by the Battelle Northwest and Lawrence Livermore Laboratories (Denham *et al.*, 1972; Lindeken *et al.*, 1972). The CaF_2:Dy energy dependence was corrected in part by enclosing the solid dosimeter in a thin two-element filter of Ta and Pb to produce a "flat" response within ± 20 percent between 50 keV and 1250 keV. Temperature dependence was determined by an experiment involving single exposures to 20 mR at temperatures ranging from $-18°C$ to 65°C. The shielded dosimeter response at room temperature decreased by about 6 percent between the first day after irradiation and the twenty-eighth day, after a very large initial amount of fading was observed. Properly packaged CaF_2:Dy is suitable for environmental radiation dosimetry, if high temperatures persist for only small fractions of the year.

Lindeken *et al.* (1972) and Becker (1972b) examined long-term fading in CaF_2:Dy that occurs after the one-day delay normally observed before readout. Dosimeters exposed to a calibration source midway through the field exposure period were assumed to provide a calibration for field dosimeters that accounts for fading. Results from this procedure may contain unevaluated errors, if the temperatures for the field and laboratory dosimeters are not comparable, if the environmental exposure rate is not essentially constant, or if the time before readout is long.

$CaSO_4$:Mn and $CaSO_4$:Dy have received much attention because of outstanding sensitivity. Lippert and Mejdahl (1967) determined that $CaSO_4$:Mn powder detected 5 μR of high energy gamma rays, while their comparable LiF detection limit was 0.5 mR, both with standard deviations of about 10 percent. The very rapid fading of $CaSO_4$:Mn thermoluminescence, however, renders this material essentially useless for even short term *in situ* environmental radiation exposures.

$CaSO_4$:Dy is a much more attractive material, because fading during the first month after irradiation is less than 2 percent, and low exposures are measurable (Yamashita *et al.*, 1971; Becker, 1972b). $CaSO_4$:Tm is similar to $CaSO_4$:Dy and its negligible neutron response may be an advantage in some applications. Although the material when used as powder requires extra handling, it is relatively inexpensive.

There may be other suitable materials and different forms of each material than those shown in Table 4-4. The materials indicated in this table have been employed in measurement programs with varying success, and illustrate the present technology.

Radiophotoluminescence (RPL) glass dosimeters may also be suitable for environmental measurements and their lower sensitivity allows exposure times of up to a year (Becker, 1972a). For example, an extensive facilities monitoring effort at Karlsruhe by Piesch (1968) indicated that changes in the expected annual exposure of about 10 mR can be considered significant. Although this type of monitoring may be useful for documenting the total exposure in facilities environs, it requires a large number of monitoring locations to establish appropriate frequency distributions at various distances to show that any increase in annual exposure is attributable to the particular facility.

4.4.3 *Suggestions for Facilities Monitoring*

Monitoring programs ought to rely on a well tested phosphor. The *in situ* exposure time can be arbitrarily selected but it should be sufficiently long to achieve a desired accuracy, say ±5 percent. The decision on exposure time will depend partly on cost and logistics, but

TABLE 4-4—*Characteristics of thermoluminescent phosphors suitable for environmental radiation measurements*

Phosphor	Z_{eff}	Room temperature fading	Self-irradiation	Reported lower exposure limit[b]	Reference[c]
		percent/month	μR H^{-1}	mR	
LiF (TLD-700)[a]	8.2	negligible	negligible	0.85	1
$Li_2B_4O_7$:Mn	7.4	≤10	none reported	—	2, 3
CaF_2:Mn	16.3	5	7–21	1.1	4, 5
CaF_2 (Natural)	16.3	negligible	9–13.8	<1	6, 7
CaF_2:Dy (TLD-200)[a]	16.3	(after preread-out anneal) negligible	negligible	0.5	8
$CaSO_4$:Dy	15.5	<2	none reported	0.5	9

[a] TLD-100, 200 and 700 are Harshaw Chemical Co. designations, but are widely used. TLD-100 (natural LiF composition) and TLD-700 (enriched in ^{7}Li) have about the same gamma-ray responses.

[b] Lower limits highly dependent on photomultiplier tube used.

[c] References: 1—Shambon (1972); 2—Binder *et al.* (1968); 3—Becker (1973); 4—Burke (1972); 5—Brinck *et al.* (1975); 6—Aitken (1968); 7—Schulman (1967); 8—Denham *et al.* (1972); 9—Yamashita *et al.* (1971).

one month is reasonable. TLD, or any dosimeter that measures the total exposure, cannot always distinguish small contributions from a nuclear facility from the variable background, so one must rely on supplementary information.

Statistical analyses may aid in evaluating the significance of apparent differences in environmental radiation at different locations for the same exposure period or changes at given locations from period to period (Piesch, 1968; Winter, 1971). Similarly, Shapiro *et al.* (1968) estimated the number of persons to be included in a sample from a given geographical area and the number of dosimeters required by each person to determine significant differences in the exposure of population groups to environmental radiation. This work provides statistical guidance for such comparisons.

Environmental measurements with RPL glass, LiF (TLD-100) and natural CaF_2 TL dosimeters, and $CaSO_4$ thermally-stimulated electron emission (TSEE) detectors, were described by Winter (1971). Annual exposures inferred from semi-annual exposure periods at many locations up to three kilometers from a reactor showed that the RPL and LiF measurements near the site boundary were about three times the expected background radiation levels. A 20 percent discrepancy between the RPL and LiF results was attributed to radiothermoluminescence fading in the glass and to the additional beta response of the unencapsulated LiF. The apparent three-fold overestimate of the CaF_2 dosimeters was attributed to the energy dependence of these unshielded dosimeters. The work indicated that the TSEE dosimeters were not suitable for lengthy exposure periods.

While the results give a qualitative indication of exposure contributions from the reactor plume, they are ambiguous and illustrate the difficulty in employing TL dosimeter results for measuring exposures near background when they are not corrected for fading, self-irradiation and energy dependence.

Great care is necessary in relying on statistical analyses, which depend on representative sampling. Near nuclear facilities or any relatively discrete source superimposed on background, one may not have the expected distribution. For a relatively large number of nuclear facilities, within the same large region, however, one can devise a monitoring program capable of detecting changes in background with time at some significant level. The difficulty of uniquely correlating such changes with a particular facility would still remain and one should, therefore, resort to supplementary information.

Another problem in the use of TLD and many other detectors is the distinction of any contribution to the radiation field by a nearby nuclear facility from other, usually natural, larger and variable contributions. EPA (1972) identified methods for accounting for these background contributions. Burke (1975) evaluated the significant errors associated with the various background subtraction methods. A semi-empirical method for estimating the expected exposure rate in a particular location, based on detailed soil moisture considerations, was described by Burke and Marcin (1974). The method was used by Burke (1975) to estimate the excess contribution to the total exposure rate to be attributed to the nuclear facility airborne effluent. It was estimated that, when the excess was in the 1–10 mR per year range above typical background, the overall uncertainty was about 1 mR per year.

TLD packaging should be light-tight and moisture proof, because many phosphors are light sensitive and hygroscopic. The dosimeters should also be protected from heat when employed in places having temperatures above about 25°C during the exposure period.

The environmental installation should be relatively secure from damage or theft, although this is difficult when large numbers of dosimeters are used in areas near nuclear facilities. In the case of nuclear facility monitoring, locations on electrical transmission line rights-of-way and government land, e.g., police and fire stations, can be conveniently arranged. Any location should be available for an indefinite period and construction areas should be avoided. Construction activities and farming can introduce measureable environmental radiation changes due to the rearrangement of natural radioactivity.

Dosimeters should be exposed in free air, away from building structures, if possible, when one wishes to monitor the contribution from nuclear facility plumes. Attaching dosimeters to posts or tree trunks is not entirely satisfactory, because of the local shielding effect, although placing them in trees is convenient and provides concealment. Winter (1971) used well-anchored aluminum tubes and Becker (1972b) used bamboo poles on Taiwan.

4.5 Gamma-Ray Spectrometry

4.5.1 *Theory*

Environmental radionuclides are identified and their contributions to total exposure rate determined by means of *in situ* gamma-ray spectrometry (Beck *et al.*, 1966: Adams and Gasparini, 1970; Minato and Kawano, 1970; Kogan *et al.*, 1969; Beck *et al.*, 1972b). Photon flux densities due to an individual radionuclide represented by the areas of total absorption peaks in the energy spectrum can be related to the concentration in an idealized environmental source and to its exposure rate contribution. Correlations rely on models of radionuclide distribution in the ground, one of the most important sources, and the consequent energy and angle spectra at various heights above the air-ground interface (Beck and de Planque, 1968; Minato, 1971; Beck *et al.*, 1972b).

An uncollimated detector above the ground in effect samples a very large volume of soil and comparable counting statistics are obtained in a small fraction of the time required for the laboratory spectrometric analysis of a collected sample. *In situ* measurements of the natural emitters, ^{40}K, ^{238}U and ^{232}Th, can be made in as few as 10 minutes with a nominal 10 × 10 cm NaI(Tl) detector, while a comparable laboratory analysis would require several hours. Collimated detectors are sometimes employed to reduce the effect of scattered gamma rays when radionuclide concentrations in rock formations are being determined (Adams and Gasparini, 1970).

A limitation of *in situ* spectrometry is that the accuracy of determinations of the exposure rate above the ground or radionuclide concentration in the ground depends on the source distribution and composition (Beck *et al.*, 1972b). Determinations of radionuclide concentrations depend strongly on the distribution with soil depth, while exposure rate determinations are much less dependent on this distribution (Beck, 1972b).

The detector response in terms of total counts in an absorption peak from a given uncollided photon flux density is determined by calibrating the detector with standard point sources. The analysis of an *in situ* spectrum and its relation to exposure rate or concentration depends on the following parameters:

(N_0/φ) = the count rate in a particular photopeak per unit flux density of photons of energy, E, incident along the detector axis of symmetry;

(N_f/N_0) = the ratio of photopeak count rate in a field situation, where the *in situ* gamma rays are not necessarily incident parallel to the detector axis of symmetry, to the count rate from an equal flux density along the detector axis. For a uniform angular response N_f/N_0 is unity. This ratio depends on the source energy and distribution, and soil properties;

(φ/C) = total uncollided flux density at the detector per unit soil concentration (pCi g^{-1}) or (mCi km^{-2}) of a particular nuclide as a function of source energy and distribution, and soil properties;

$(\dot{X})$ = Exposure rate in μR h^{-1} at one meter above the ground from all photons from a particular nuclide including the secondary photons produced in the soil and air; and

$(\varphi/\dot{X})$ = the ratio of the flux density at the detector due to photons of energy, E, from the decay of a particular nuclide and any daughters to the corresponding total exposure rate for that nuclide and its daughters.

Then the photopeak count rate is related to the exposure rate in air above the ground by:

$$(N_f/\dot{X}) = (N_f/N_0)(N_0/\varphi)(\varphi/\dot{X}) \text{ min}^{-1}/\mu\text{R h}^{-1} \quad (4\text{-}6)$$

and, similarly, to radionuclide concentrations in the ground by

$$(N_f/C) = (N_f/N_0)(N_0/\varphi)(\varphi/C) \text{ min}^{-1}/\text{pCi g}^{-1} \text{ or mCi km}^{-2}. \quad (4\text{-}7)$$

The high resolution of Ge(Li) detectors allows one to measure absorption peak areas with little interference from neighboring peaks and several peaks from the same nuclide. Ge(Li) detector counting efficiencies are low and somewhat longer *in situ* counting times are required to obtain statistical precision comparable to that of NaI(T1).

The first two terms in Eqs. 4-6 and 4-7 represent the angular response correction and counting efficiency of the particular detector, respectively. The ratios $\varphi/\dot{X}$ and φ/C depend only on source composition and geometry and can be used for any detector. The values for $\varphi/\dot{X}$ in Tables 4-5 and 4-6 are inferred from Tables 2-15 and 2-17 and decay scheme data in Tables 2-1 through 2-5.

4.5.2 *Exposure Rate Measurements*

The $\varphi/\dot{X}$ for common radionuclides, useful for the calibration of gamma-ray spectrometers, are listed in Tables 4-5 and 4-6. It should be noted that the principal natural contributors to exposure rate one meter above the ground are ^{214}Bi, ^{208}Tl, and ^{40}K, as shown in the major peaks of a typical field spectrum (Figure 4-2) obtained in Connecticut in 1971.

The determination of the detector energy response (N_0/φ) and angular response (N_f/N_0) is accomplished in the laboratory by measuring known uncollided flux densities incident either parallel or at known angles to the detector axis of symmetry. Corrections for air attenuation and source self-absorption may be necessary. NaI(Tl) detectors should be calibrated with source energies simulating the prominent peaks expected in field measurements. Difficulties in ob-

TABLE 4-5—*Values of $\phi/\dot{X}$ one meter above soil containing uniformly distributed naturally-occurring radionuclides*[a]

Parent Nuclide	Energy	$\phi/\dot{X}$	Parent Nuclide	Energy	$\phi/\dot{X}$
Uranium Series	keV	$cm^{-2}\ s^{-1}/\mu R\ h^{-1}$	Thorium Series	keV	$cm^{-2}\ s^{-1}/\mu R\ h^{-1}$
^{226}Ra	186	2.52(−3)[b]	^{228}Ac	129	1.03(−3)
^{214}Pb	242	5.71(−3)		210	2.06(−3)
	295	1.60(−2)	^{212}Pb, ^{224}Ra	239, 241	2.57(−2)
	352	3.30(−2)			
^{214}Bi	609, 666	5.37(−2)	^{228}Ac, ^{208}Tl, ^{228}Ac	270, 277, 282	3.62(−3)
	768	6.43(−3)			
	934	4.45(−3)	^{212}Pb	301	1.96(−3)
	1120	2.31(−2)	^{228}Ac	338	7.73(−3)
	1238	9.45(−3)	Mixed	328–340	1.03(−2)
	1378	8.19(−3)	^{228}Ac	463	3.26(−3)
	1401–08	6.87(−3)	^{208}Tl	510	6.84(−3)
	1510	3.91(−3)	^{228}Tl	583	2.27(−2)
	1730, 1765, 1845	3.95(−2)	^{212}Bi, ^{228}Ac	727	6.60(−3)
			^{228}Ac	755	9.57(−4)
				772	1.45(−3)
	2205	1.07(−2)		795	4.25(−3)
	2448	3.66(−3)		830, 835, 840	3.33(−3)
Potassium			^{208}Tl	860	4.18(−3)
^{40}K	1464	0.203	^{228}Ac	911	2.68(−2)
				965, 969	2.17(−2)
				1588	4.36(−3)
			^{208}Tl	2615	5.92(−2)

[a] Taken from Beck *et al.* (1972b). For the uranium and thorium series in equilibrium, the total exposure rate per unit concentration is 1.82 $\mu R\ h^{-1}/pCi\ g^{-1}$ and 2.82 $\mu R\ h^{-1}/pCi\ g^{-1}$, respectively. For potassium, this value is 0.179 $\mu R\ h^{-1}/pCi\ g^{-1}$.

[b] $2.52(-3) = 2.52 \times 10^{-3}$.

taining a calibrated ^{40}K source are overcome by resorting to ^{24}Na (1.37 MeV) or ^{42}K (1.52 MeV). Cross calibrations of NaI(Tl) detectors with the more accurate Ge(Li) are helpful. Ge(Li) calibrations require the use of many sources.

The measured peak areas in NaI(Tl) do not usually account for all the totally absorbed gamma rays, because the continuum below the peak is usually so large. However, the shapes of NaI(Tl) continua in field spectra are relatively constant because the Compton scattered gamma rays from natural emitters in the soil dominate the NaI(T1) spectrum. Thus, if one calibrates the unshielded detector in a situation similar to that in the field (e.g. including laboratory background), the photopeak count rate fraction obtained is essentially equal to the fraction obtained in the field for the same incident flux density.

TABLE 4-6—*Calculated $\phi/\dot{X}$ one meter above ground containing various distributions of manmade radionuclides*[a]

Isotope	Energy	Source distribution (α/ρ in $cm^2\ g^{-1}$)					
		0.0625	0.206	0.312	0.625	6.25	∞ (Plane)
	keV	$cm^{-2}s^{-1}/\mu Rh^{-1}$	$cm^{-2}s^{-1}/\mu Rh^{-1}$	$cm^{-2}s^{-1}/\mu Rh^{-1}$	$cm^{-2}s^{-1}/\mu Rh^{-1}$	$cm^{-2}s^{-1}/\mu Rh^{-1}$	$cm^{-2}s^{-1}/\mu Rh^{-1}$
^{144}Ce	134	1.04	1.19	1.31	1.48	2.14	1.97
^{144}Ce-^{144}Pr	134	0.352	0.453	0.504	0.577	0.838	0.780
^{141}Ce	145	1.17	1.42	1.52	1.74	2.45	2.23
^{131}I	364	0.444	0.541	0.585	0.636	0.808	0.702
^{125}Sb	428	0.149	0.177	0.193	0.203	0.259	0.245
^{140}La	487	0.046	0.056	0.061	0.064	0.083	0.076
^{140}Ba-^{140}La	487	0.037	0.045	0.049	0.052	0.067	0.062
^{103}Ru	497	0.416	0.503	0.528	0.574	0.715	0.687
^{106}Ru-^{106}Rh	512	0.251	0.303	0.319	0.339	0.419	0.401
^{140}Ba	537	0.296	0.352	0.370	0.400	0.486	0.465
^{140}Ba-^{140}La	537	0.0206	0.0253	0.0268	0.0282	0.0362	0.0334
^{120}Sb	601	0.104	0.121	0.128	0.135	0.166	0.160
^{103}Ru	610	0.0271	0.0322	0.0335	0.0359	0.0440	0.0430
^{106}Ru	622	0.129	0.153	0.160	0.170	0.207	0.198
^{137}Cs	662	0.378	0.441	0.463	0.499	0.606	0.582
^{95}Zr	724	0.155	0.181	0.192	0.206	0.247	0.241
^{95}Zr-^{95}Nb	724	0.0476	0.0557	0.0589	0.0634	0.0758	0.0734
^{95}Zr	757	0.196	0.230	0.242	0.265	0.313	0.303
^{95}Zr-^{95}Nb	757	0.0602	0.0709	0.0744	0.0815	0.961	0.0922
^{95}Nb	766	0.346	0.409	0.428	0.473	0.551	0.531
^{95}Zr-^{95}Nb	766	0.239	0.282	0.297	0.328	0.381	0.371
^{140}La	816	0.0287	0.0336	0.355	0.0374	0.0457	0.0433
^{140}Ba-^{140}La	815	0.0232	0.0270	0.0286	0.0302	0.0368	0.0350
^{54}Mn	835	0.332	0.380	0.400	0.438	0.509	0.497
^{140}La	1597	0.154	0.166	0.173	0.175	0.203	0.195
^{140}Ba-^{140}La	1597	0.124	0.134	0.140	0.142	0.163	0.158
^{60}Co	1173	0.127	0.146	0.153	0.162	0.188	0.183
^{60}Co	1333	0.134	0.151	0.159	0.168	0.191	0.187

[a] From Beck *et al.* (1972b).

Photopeak areas are determined by fitting the continua in some fashion, often with an exponential function, and ascribing the counts represented by the area of the peak above the fitted curve to the incident uncollided flux density. Beck *et al.* (1972b) showed that the statistical error in an *in situ* 30-minute counting time with a 60-cm^3 coaxial Ge(Li) detector was more significant than differences due to methods of fitting the continuum beneath the strongest peaks. Determining peak area count rate is fairly easily done with a simple computer code, if many spectra are to be analyzed, or graphically from a printout of the pulse height spectrum.

The angular response correction, N_f/N_0, is determined by measuring monoenergetic sources located at various angles from the detector

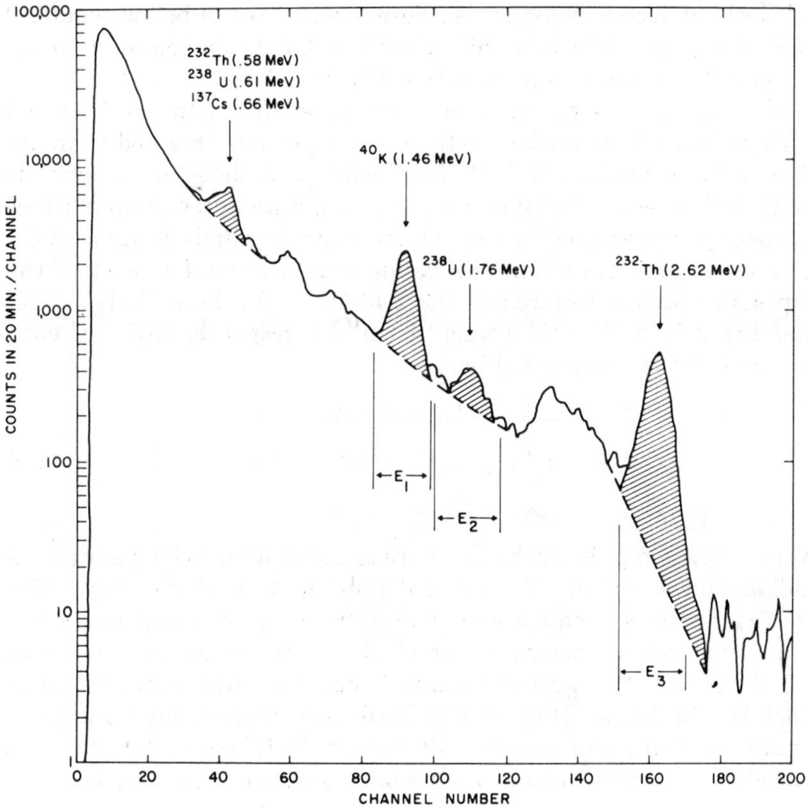

Fig. 4-2. *In situ* NaI(Tl) spectrum of background and fallout radiation in Connecticut in 1971.

axis. Usually the detector has azimuthal symmetry. If $N(\theta)/N_0$ is the ratio of the response for an energy E and an angle θ with respect to the detector axis, then

$$\frac{N_f}{N_0} = \frac{1}{\varphi} \int_0^{\pi/2} \varphi(\theta) \frac{N(\theta)}{N_0} \, d\theta \tag{4-8}$$

The N_f/N_0 values are determined by numerically integrating the equation, using a smooth fit to measured angular response data to interpolate over $N(\theta)/N_0$. Because N_f/N_0 is usually nearly unity, interpolation errors are unimportant (Minato, 1971; Beck *et al.*, 1972b). Many detectors have reasonably isotropic responses; a 10 × 10-cm NaI(Tl) detector response is probably within a few percent of unity (i.e., 10 percent or less) except for energies below the most prominent natural contributors to exposure rate. For a right cylindrical Ge(Li) detector, N_f/N_0 is also close to unity except below about 200 keV. Elongated detectors will, of course, have complex angular responses that require determination (Anspaugh *et al.*, 1972).

When naturally-occurring nuclides associated with the ^{238}U and ^{232}Th series and ^{40}K predominate, an "energy band" method of analysis can be applied to NaI(Tl) field spectra. A detector response is computed in terms of counts per pulse height analyzer channel times the energy represented by each channel for the bands designated E_1, E_2 and E_3 in Figure 4-2. The bands that account for most of the exposure rate are centered on the 1.46 MeV ^{40}K, the 1.76 MeV ^{214}Bi, and the 2.62 MeV ^{208}Tl peaks. Then the response, $R(E)$, of each energy band is represented by:

$$\begin{aligned} R(E_1) &= \mathrm{u}_1\dot{X}_\mathrm{U} + k_1\dot{X}_\mathrm{K} + \mathrm{t}_1\dot{X}_\mathrm{T} + \mathrm{C}_1 \\ R(E_2) &= \mathrm{u}_2\dot{X}_\mathrm{U} + \mathrm{k}_2\dot{X}_\mathrm{K} + \mathrm{t}_2\dot{X}_\mathrm{T} + \mathrm{C}_2 \\ R(E_3) &= \mathrm{t}_3\dot{X}_\mathrm{T} + \mathrm{C}_3 \end{aligned} \tag{4-9}$$

where the $\dot{X}$'s represent the exposure rates from the principal natural radionuclides and the C's the contributions from cosmic rays. The coefficients, u, k, and t are evaluated from regression analyses of a number of *in situ* spectra for which the partial exposure rates from ^{40}K, ^{214}Bi and ^{208}Tl were determined independently from photopeak analyses. Beck *et al.* (1966 and 1972b) developed the following approximate equations for a nominal 10 × 10 cm NaI(Tl) detector, which is uncollimated and shielded against beta radiation with bakelite:

$$\begin{aligned} \dot{X}_\mathrm{u} &= 0.4\, R'(E_2) - 0.2\, R'(\mathrm{E}_3) \\ \dot{X}_\mathrm{T} &= 0.3\, R'(E_3) \quad \text{and} \\ \dot{X}_\mathrm{K} &= 0.08\, R'(E_1) - 0.06\, R'(E_2) - 0.02\, R'(E_3) \end{aligned} \tag{4-10}$$

where the $R'(E)$'s, in this case, are in units of GeV per 20 minute counting period. The primes in the $R(E)$'s indicate that the cosmic-ray contributions to $R(E)$ must be subtracted before calculating the exposure rates (Beck *et al.*, 1972b). The specified energy bands are:

$$\begin{aligned} &1.32 \text{ MeV} < E_1 < 1.60 \text{ MeV} \\ &1.62 \text{ MeV} < E_2 < 1.90 \text{ MeV, and} \\ &2.48 \text{ MeV} < E_3 < 2.75 \text{ MeV.} \end{aligned}$$

The exposure rate from natural sources in the ground, as obtained from an *in situ* 10 × 10 cm NaI(Tl) spectrum between 0.15 and 3.4 MeV, is approximately

$$\dot{X}_\mathrm{t} = R'\ (E_\mathrm{t})/37 \tag{4-11}$$

where $R'(E_t)$ is the response in GeV/20 minutes counting period within the indicated energy range (Beck *et al.*, 1972b). This method provides a reasonable measure of the exposure rate because the *in situ* spectrum above a few hundred keV in air due to natural emitters does not depend sensitively on the exact proportions of uranium, thorium, and potassium. The mass energy absorption coefficient is fairly constant for many materials (e.g., ICRU 1964) and calculations indicate that most of the exposure rate is due to the energy range between about 0.1 and 1.5 MeV (Beck and de Planque, 1968; Moriuchi and Miyanaga, 1966). This finding also applies to scintillation survey meters (Section 4.3).

Confidence in measurements is improved by comparing environmental exposure rates determined by the photopeak and energy band methods of *in situ* spectrometry with independent ionization chamber measurements.

In general, the largest percentage error in exposure rate is obtained for the ^{238}U series, because of radon movement that produces a somewhat altered source distribution and the difficulty in measuring the small flux density of 1.76 MeV (^{214}Bi) gamma rays from the NaI(T1) absorption peak. Beck *et al.* (1972b) estimated the ^{238}U series exposure rates to be correct to about 25 percent. Because of the ability to resolve the 295-, 352- and 609-keV U peaks with a Ge(Li) detector, one obtains ^{238}U exposure rate measurements estimated to have an accuracy of better than 15 percent. The most accurate Ge(Li) measurements of exposure rate from ^{40}K and thorium are correct to about 10 percent (Beck *et al.*, 1972b). The uncertainty in ^{137}Cs exposure rates from deposited fallout is ± 15 percent with the NaI(Tl) detector and $\pm$ 10 percent with the Ge(Li) under most circumstances.

It should be noted that the determination of exposure rates from particular radionuclides in the airborne effluent of a nuclear facility depends on the source distribution of the plume, which varies rapidly with time. Additional research is required to establish ranges of validity of such measurements. However, spectrometric data, such as those in Figure 4-3, can be used to apportion total exposure rate measurements to the principal contributors. This figure shows a Ge(Li) spectrum under a noble gas plume as a dashed line and with the plume absent as a solid line (Beck *et al.*, 1971). The background and fallout peaks are not labeled.

Gold *et al.* (1973) developed a method in which the continuum is determined with a low atomic number detector having a response primarily dependent on Compton interactions. The gamma-ray field is not significantly perturbed and any spectral unfolding is simplified. They report that the method is essentially one of absolute

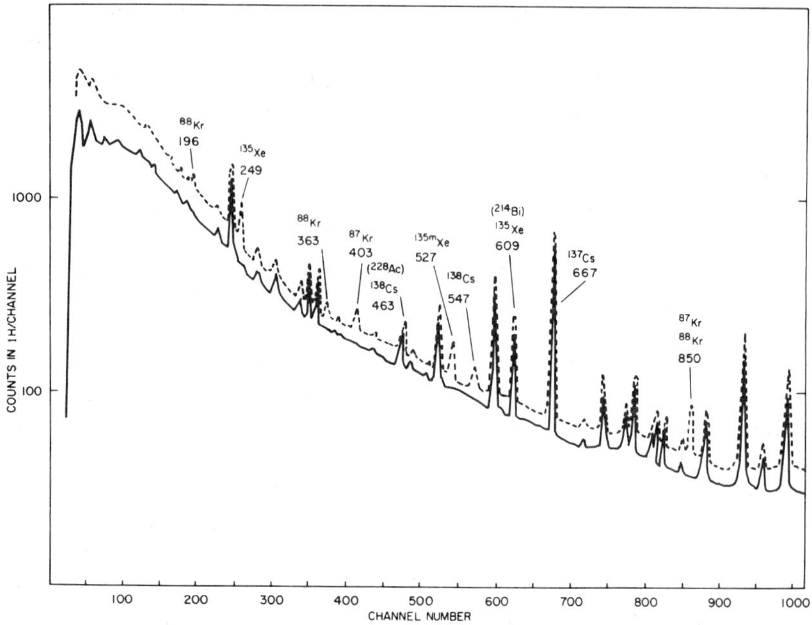

Fig. 4-3. *In situ* Ge(Li) spectra near a nuclear power reactor. (Dashed spectrum includes noble gas emitters; solid spectrum is normal background; selected nuclides are labeled for reference.) [From Beck, H. L., Lowder, W. M. and McLaughlin, J. E. (1971). "*In situ* external environmental gamma-ray measurements utilizing Ge(Li) and NaI(Tl) spectrometry and ionization chambers," page 499 in *Rapid Methods for Measuring Radioactivity in the Environment* (International Atomic Energy Agency, Vienna), by permission].

dosimetry and does not require independent measurements. Limitations include the long measurement times for statistically significant results. The overall error was estimated to be about 20 percent, but somewhat greater near 2 MeV. More recent measurements with a 10-cm diameter sphere of liquid xylene allow the simultaneous determination of the cosmic-ray absorbed dose rate from muons (Gold *et al.*, 1974).

4.5.3 *Radionuclide Concentration Measurements*

The determination of radionuclide concentrations in the environment by *in situ* spectrometry has been employed in geochemical and geophysical surveys (Adams and Fryer, 1964; Mahdavi, 1964; Doig, 1968; Kogan *et al.*, 1969; and many others) and in special studies for determining concentrations of natural and manmade nuclides in the ground (Beck *et al.*, 1964 and 1972b; Anspaugh *et al.*, 1972).

Adams and Fryer (1964) compared hundreds of laboratory analyses of collected rock samples with *in situ* measurements using a collimated NaI(Tl) detector. This instrumentation was capable of determining ^{214}Bi and ^{40}K activity within a few percent, although the uncertainty of the ^{238}U concentration inferred from the 1.76 MeV photopeak analysis depends on the degree of secular equilibrium. Such *in situ* systems sample about one hundred times the sample weights used for laboratory analyses (Mahdavi, 1964).

Laboratory analyses of soil samples for naturally occurring radioactivity by the Lawrence Berkeley Laboratory, Argonne National Laboratory, and Rice University were compared with *in situ* NaI(Tl) spectrometric measurements made in 1963 by Beck *et al.* (1964). Conversion of measured collected soil sample concentrations to exposure rates was made using the factors in Table 2-18. The ^{40}K and thorium results generally agreed within ±25 percent but uranium series concentration determinations agreed less well. Systematic differences in these measurements are attributed to the usual laboratory procedure of drying samples and re-establishing $^{226}Ra - {}^{222}Rn$ equilibrium before counting. Additional comparisons by Beck *et al.* (1972b) indicated that agreement for ^{137}Cs soil activity was within a factor of two when the vertical distribution was unknown and an assumed value for φ/C was used in Eq. 4-7, and within ±20 percent when the vertical distribution was determined.

Anspaugh *et al.* (1972) employed an unshielded 70-cm^3 closed-end coaxial Ge(Li) detector for *in situ* measurements in the Livermore Valley area calibrated by the method indicated in Section 4.5.1. Soil samples were collected in increments to a depth of 25 cm to determine the vertical distribution (α/ρ) and the total activity or the radionuclides. The results from one location used for field calibration, shown in Figure 4-4, indicate that natural radioactivity is distributed uniformly and ^{137}Cs from world-wide fallout decreases exponentially with depth. The average concentrations for the thorium and uranium series and for ^{40}K and ^{137}Cs from the *in situ* measurements and four separate laboratory analyses agreed within a few percent. In cases where α/ρ was not measured, agreement was within 15 percent for the natural, uniformly distributed nuclides, and within a factor of 2 for ^{137}Cs.

Additional work is desirable to facilitate basic geophysical studies and monitoring long-term changes in manmade radionuclide concentrations. Other applications of *in situ* spectrometry, concerning the detection and quantification of manmade nuclides in the sediments of water bodies, the detection and measurement of plutonium and other transuranic elements in the environment, and aerial surveying, are described below.

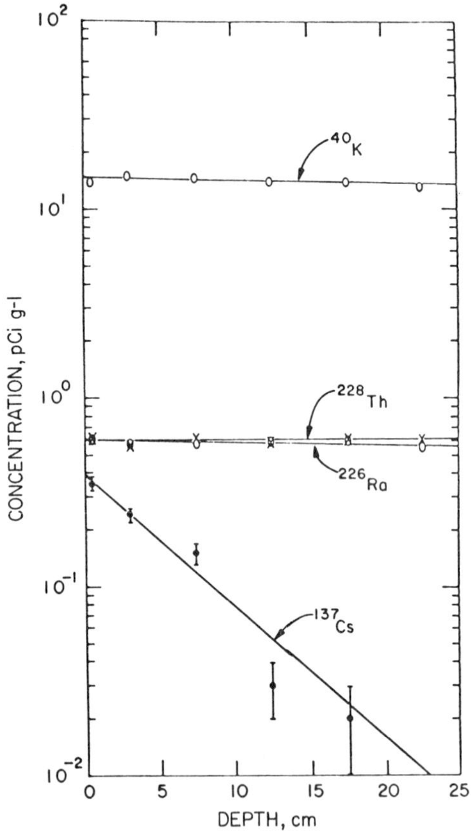

Fig. 4-4. Soil depth profiles near Livermore, California (from Anspaugh *et al.*, 1972).

4.5.4 *Measurement of Low-Energy Photons from Environmental Plutonium*[4]

The appearance of locally and globally distributed plutonium in the environment has indicated a continuing need for convenient and reliable measurement methods. Plutonium usually comprises a complex mixture of ^{239}Pu, ^{240}Pu, ^{241}Pu and ^{242}Pu, and possibly ^{238}Pu plus ^{241}Am from the decay of ^{241}Pu (Gunnink and Morrow, 1971). Plutonium presently in the environment is mostly ^{239}Pu, although material in nuclear power production ranges up to about 40 percent by weight of both ^{239}Pu and ^{240}Pu, most of the alpha activity being due to ^{240}Pu and

[4] This section concerns the possible *in situ* detection of ^{239}Pu and ^{241}Am, although other transuranic nuclides may become problems. *In situ* techniques are needed to supplement current sample collection and laboratory analysis methods.

most of the beta activity to ^{241}Pu. Dispersion of plutonium has resulted from nuclear weapons tests, atmospheric burnup of a ^{238}Pu auxiliary nuclear power source, and local dispersion from nuclear facilities. Determining and controlling additions of plutonium are important in view of present efforts toward developing nuclear power from plutonium.

The most reliable and sensitive methods are based on sample collection and laboratory analysis, as indicated in Section 6, and are capable of determining plutonium deposits from fallout below 1 mCi km^{-2} (Hardy *et al.*, 1973). Determining environmental plutonium levels by sampling and analysis is tedious and costly, and there is a need for supplemental *in situ* measurement methods. The same rationale led to the development and use of high-energy gamma-ray spectrometry in geological studies.

Current *in situ* measurements of plutonium have been used primarily to determine relatively rapidly the extent of fairly recent "spills", where the plutonium deposit is near the ground surface (Healy, 1971).

Measurements of 60-keV photons which accompany about 36 percent of ^{241}Am disintegrations, and the concurrently emitted photons at 17 keV from the excited daughter ^{237}Np, have provided the *in situ* detection mechanism for these relatively large concentrations of plutonium. However, the inference of concentrations from such measurements depends on knowing the nuclide composition and the distribution in the ground. For example, the 60-keV flux density from ^{241}Am distributed in an ideal plane is many times that from the same amount of ^{241}Am distributed deeply (see 50-keV data in Table 2-15, $\alpha/\rho = 0.063$ cm^2 g^{-1}). *In situ* measurements may be further complicated by interference from the decay of nuclides in the uranium and thorium series which results in the emission of de-excitation x rays that may be erroneously attributed to ^{241}Am.

Tinney *et al.* (1969) and Schmidt and Koch (1966) described measurements utilizing a 12.5-cm diameter, thin NaI(Tl) detector to measure freshly deposited plutonium. This instrument, the FIDLER[5], records the photon flux densities in energy bands centered on the 17 and 60 keV photopeaks shown in the spectrum of Figure 4-5 from a ^{241}Am calibration source. The peak near 17 keV in the background spectrum is attributable to naturally-occurring ^{235}U.

Calibration with plane sources is impractical, because they are difficult to fabricate, maintain, and use, and suitable environmental plutonium sources are fortunately rare. Calibration with point sources, as in the case of any *in situ* gamma-ray spectrometer, must

[5] Acronym for Field Instrument for Detection of Low Energy Radiation.

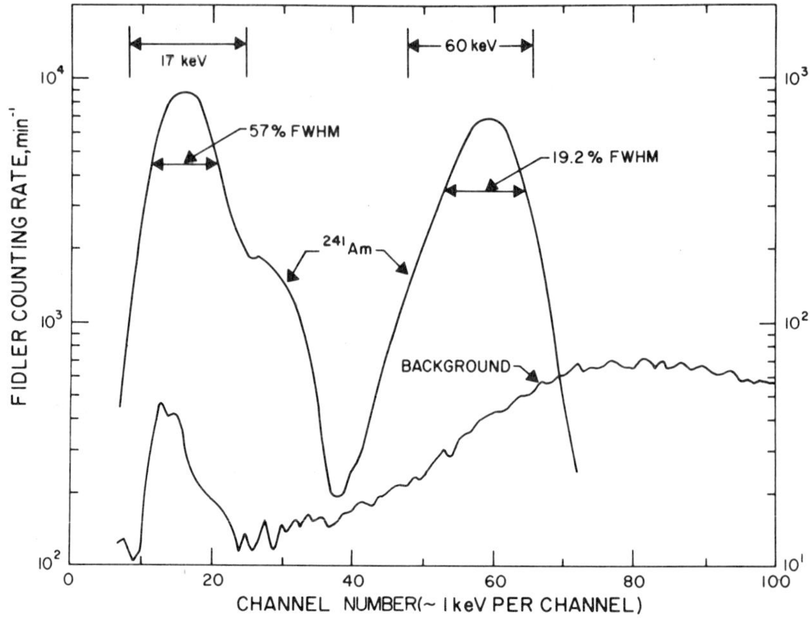

Fig. 4-5. FIDLER response to background (right ordinate) and to ^{241}Am calibration source (left ordinate) (from Tinney *et al.*, 1969).

be properly transformed to the required distributed source.

The point calibration source locations should be along a horizontal radius from the vertical axis of the detector, positioned at a convenient height, and spaced according to equal angle increments with the detector. The number of data points suitable for determining the response to an area source can be estimated from standard formulations for this pseudo line source (Goussev *et al.*, 1968). Tinney *et al.* (1969) have shown that 10° increments for point source locations lead to a suitable calibration and the contributions from sources located greater than about two meters contribute little to the plane source response for small detector heights. For heights above two meters, the maximum radical displacement of the calibration source should be greater.

The reported sensitivity for the NaI(Tl) detector measurements in the 17-keV ^{237}Np x-ray energy band is based on calibration with an ^{241}Am source at a detector height of 30.5 cm and a projected plane source area of 25 m^2 (Tinney *et al.*, 1969). Provided that interference from background is either negligible or fully accounted for, the minimum detectable ^{241}Am sensitivity at a 95 percent confidence level was estimated to be about 20 nCi m^{-2} (Lindeken and Hodgkins, 1969). In practice one cannot expect to attribute the 17-keV photon counting rate uniquely to ^{239}Pu, because of the complex isotopic composition of the mixture and the contribution of background x rays must be evaluated (Lindeken and Hodgkins, 1969). Measurements of the 60-keV ^{241}Am gamma-ray flux density also require prior knowledge of the plutonium isotopic composition and nuclides that interfere, although detection of the more energetic photon somewhat relaxes the measurement problem.

Hiebert *et al.* (1973) and previously Tinney *et al.* (1969) evaluated the spectral output of thin NaI(Tl) detectors in order to account for background radiation. This method was reported to have a detection limit of about 100 nCi m^{-2} for ^{239}Pu and 10 nCi m^{-2} for ^{241}Am distributed in a plane for areas having well known backgrounds.

The technique of comparing the low-energy photon detector counting rate observed in locations where the plutonium concentrations have been determined from soil samples (Healy, 1971) may remove some of the uncertainty in inferring concentrations in other locations, but the depth distributions must be known (see Section 4.5.3). Even so, correlations between FIDLER-type measurements and soil sample measurements are often poor, unless the plutonium deposit is recent (Lindeken, 1973).

If, as expected, there is to be use of plutonium, it is obviously desirable to develop much improved *in situ* measurement capabilities. A desirable design aim is the rapid, reliable detection of concentrations of ^{241}Am and ^{239}Pu of the order of 10 nCi m^{-2}. Work with high resolution intrinsic germanium detectors is aimed at this application (Armantrout *et al.*, 1974). Another possibility, field alpha counting of carefully collected soil samples, is described in Section 4.7.2.

Thin phoswich detectors, similar to those employed for the *in vivo* determination of plutonium in man, should have better resolution than the FIDLER detector and discriminate against de-excitation x rays and degraded higher energy photons. If the 17-keV photons attributable directly to ^{238}Pu and ^{239}Pu can be detected and sensitivity is adequate, a convenient method for selecting sample collection locations can be developed.

4.5.5 *Underwater Spectrometry*

Gamma-ray spectrometric methods are useful for determining natural and manmade radionuclide concentrations in water to obtain indications of valuable minerals (Miller and Symons, 1973) and to evaluate distributions of fallout from weapon tests to obtain indications of natural transport processes (Volchok *et al.*, 1970; Seymour, 1971).

Rugged underwater detectors and portable multichannel analyzers have been used by Riel (1966) and Plato and Gelsky (1972). The sensitivity of NaI(Tl) spectrometers is better than 10 pCi l^{-1}, the greatest being achieved by 10 × 10-cm or larger NaI(Tl) detectors for counting times of an hour or longer. Gelsky *et al.* (1970) used an amphibious spectrometer for measurements on land and near sediments in fresh water bodies.

A spectrometer installed in a nuclear facility discharge stream provides information on the quantity and composition of effluent radionuclides at relatively small cost (Riel *et al.*, 1973). In this application, the detector may become dirty or contaminated, and suffer damage. When several release points exist, one should monitor for small "seeps" downstream to document significant changes.

Instruments like those described in Section 4.3 can be adapted for underwater use and operated as both energy spectrometers and total intensity meters. For total intensity measurements, plastic and liquid scintillators are rugged and relatively inexpensive, compared to NaI(Tl), although their resolution is low (Nelepo, 1960).

Most measurements have been made with NaI(Tl) scintillators (Proctor *et al.*, 1962; and Chesselet *et al.*, 1965). Cesium iodide detectors may be an attractive alternate, because of greater counting efficiency, resistance to mechanical damage, and lower affinity for water.

High resolution Ge(Li) detectors, despite low efficiency and high cost, should be applied to underwater spectrometry for more precise identification of individual nuclides. One can compare the extensive NaI(Tl) experience with that anticipated with Ge(Li) detectors. The *in situ* response of the former depends on the effective surface area and the total absorption probability of the detector material (Neiler and Bell, 1965). The total counting efficiency usually given for solid state detectors is that relative to a 7.6 × 7.6-cm NaI(Tl) detector. Neiler and Bell (1965), Cline and Heath (1967), and others have published total absorption probabilities for 7.6 × 7.6-cm NaI(Tl) and for some Ge(Li) detectors.

In analyzing the net count rate for an entire spectrum, a particular energy band, or a particular absorption peak, one usually estimates the lower limit of detection (LLD) as described in a forthcoming NCRP report (NCRP, 1977) and in Pasternack and Harley (1971). To halve the LLD one must double the counting efficiency, reduce background by four, or increase counting time fourfold.

Counting efficiencies and typical background responses to gamma rays in the various energy bands for an underwater NaI(Tl) detector are given in Table 4-7 (Riel *et al.*, 1973). These values may be

TABLE 4-7—*Efficiency and background counting rate for underwater NaI(Tl) detector*

Energy[a]	Efficiency	Background counting rate[b]		
		Detector[c]	Ocean	Sediment[d]
MeV	$h^{-1}/pCi\ l^{-1}$	h^{-1}	h^{-1}	h^{-1}
0.13	20	2,500	1,000	10,000
0.3	60	2,000	2,000	30,000
0.6	90	1,500	1,500	20,000
1.0	120	1,300	1,300	20,000
1.5	150	3,000	4,000	40,000
2.0	170	40	0	4,000
2.5	190	60	0	8,000

[a] Energy window $\Delta E/E$ = 18 percent for 12.7 × 12.7-cm NaI(Tl) detector enclosed in a steel vessel with a wall thickness of 0.64 cm.

[b] Cosmic-ray contribution between 0.13 and 2.5 meV is 500 counts h^{-1} 3 m under water, 70 at 30 m and 1 at 305 m (see, for example, Riel and Simons, 1962).

[c] Mostly due to phototube; detector systems without special low background materials may more than double this.

[d] Typical, but varies greatly with K, Ra, Th content.

extrapolated to other NaI(Tl) detectors by

$$E_2/E_1 = (A_2/A_1)\ (P_2/P_1)e^{(\mu_1 x_1 - \mu_2 x_2)} \tag{4-12}$$

This equation takes account of the overall counting efficiencies (E), detector surface areas (A), photopeak efficiencies (P), and the linear absorption coefficients (μ) and the thickness (x) of the detector containers. The error in extrapolating from the 12.7 × 12.7 cm scintillator in a 0.6-cm steel wall to a 5.1 × 5.1-cm detector in a very thin wall is about 20 percent.

The efficiency of the detector in Table 4-7 for a mixture of a reactor liquid effluent is 1,000 ± 200 $h^{-1}/pCi\ l^{-1}$ for the whole spectrum above 100 keV (Riel *et al.*, 1973). The estimated lower limit of detection is 3 pCi l^{-1} for typical reactor liquid effluent, using a counting time of five minutes, and also for ^{137}Cs about 3 meters deep in the ocean, with a counting time of one hour.

Background fluctuations limit the sensitivity and must be measured or suppressed, although measurements of a few specific nuclides are less influenced by background. After normalizing the measured spectrum for energy and counting time to an appropriate calibration spectrum, background is subtracted, and the concentration of the nuclides inferred (Riel *et al.*, 1969).

Calibration spectra are derived from measurements of many monoenergetic standards. The spectrometer counting efficiency varies with the composition of radionuclides. By measuring many calibration solutions, one can determine the standard deviation of the efficiency, and thus the suitability of the total spectral count rate for a particular environmental or effluent determination. One can also calibrate in liquid effluent holdup tanks. The fully immersed detector should be at least one and preferably two meters from any wall. The measured energy spectra are normalized for energy and concentration and retained for later analysis. The variance of repeated measurements should be small if the composition is accurately known. One may also employ Monte Carlo codes to generate the response to any nuclide (Madigosky and Simons, 1966).

The simplest interpretation of a measurement is that the released radioactivity is or is not within specified limits. The main value of environmental spectrometry, however, is to identify nuclides present and determine their concentrations. This implies that gamma-ray spectrometers are to be used to analyze radionuclides in water over large ranges of concentrations and conditions. The experience from geophysical studies is now being applied to light water reactor effluent monitoring (Riel *et al.*, 1973; Kahn *et al.*, 1972) and other sources, such as hospital sewers, may also require investigation.

4.6 Airborne Radiation Surveys

4.6.1 *Historical Development*

Airborne measurements fulfill requirements for large area prospecting for ore bodies and monitoring for large radioactivity releases (Williams *et al.*, 1959). Davis and Reinhardt (1957) developed one of the first systems consisting of six NaI(Tl) detectors called Aerial Radiological Measuring System (ARMS I). A later system (Merian *et al.*, 1960) was developed for monitoring during nuclear weapons tests and subsequently it has been used for many large area surveys (Doyle, 1972).

Davis and Reinhardt (1962) refined operational techniques by means of a calibration experiment in which hundreds of ^{60}Co and ^{137}Cs point sources were distributed over a large area. Gregory and Horwood (1963) measured spectral characteristics, air distance variations and other parameters as a prelude to geophysical survey mapping in Canada.

Extensive searches have been made for uranium deposits where anomalies in the natural gamma-ray field are evaluated from uranium-to-thorium ratios as reflected in measured photopeaks of ^{214}Bi and ^{208}Tl (IAEA, 1969b). Analysis of the uranium-to-potassium ratios and total spectral count rates to determine concentrations of natural elements in soil and rocks is a valuable aid to large scale geologic mapping and exploring for a variety of mineral resources (Darnley and Grasty, 1971; Foote, 1968; Pitkin, 1968; Schwarzer *et al.*, 1972).

Large areas surrounding major nuclear facilities have been surveyed to document terrestrial radiation baselines (Doyle, 1972; Guillou, 1964), to aid in environmental surveillance (Beck, C. K., 1971) and to provide information on radiological emergency planning (Deal *et al.*, 1971).

Radiation attenuation by snow or ice and moisture in the soil is related to the equivalent water mass. Millions of square kilometers have been surveyed in the Soviet Union to obtain water equivalent snow cover contour maps for compiling hydrological forecasts (Kogan *et al.*, 1969). Studies have been done in the United States (Peck *et al.*, 1971; Fritzsche and Burson, 1972) on applying aerial survey techniques to water resources management and evaluation of average soil moisture changes over large areas (Burson and Fritzche, 1972).

4.6.2 *Instrumentation and Data Acquisition*

Both single NaI(Tl) scintillation detectors and arrays are used extensively in most airborne survey systems (Doyle, 1972; Adams, 1969; Darnley, 1970; Stuart, 1971). Foote (1969) employed six detectors as the primary array and a seventh crystal shielded with lead on the bottom for estimating the airborne ^{214}Bi contribution to the other six detectors. Dmitriev *et al.* (1971) and Kogan *et al.* (1969) described a two detector system in which one is unshielded and the other is shielded on the top and sides with lead having an angular opening on the bottom.

Large plastic scintillator systems have an advantage over NaI(Tl) for exposure rate measurements because of their lesser dependence on gamma-ray energy (Dahlstrom, 1965; Duval *et al.*, 1972). Ge(Li) detectors are promising, but NaI(Tl) will probably remain the primary detector for many years.

Scintillator counts are stored in both single and multi-channel pulse-height analyzers. The energy width of the channel is chosen to maximize the signal-to-noise ratio for detecting primary gamma rays from the nuclide of interest. Grasty and Darnley (1971) used channel widths centered on 1.46, 1.76, and 2.62 MeV as indicated in Figure 4-2. The potassium, uranium, and thorium are then inferred from measurements of the flux densities from ^{40}K, ^{214}Bi and ^{208}Tl, respectively. The total count rate from all photons above 50 keV (ARMS, 1973) at a height of 100 m is approximately proportional to exposure rate at one meter above the ground (see Section 2.3.3). Kogan *et al.* (1969), Dmitriev *et al.* (1971), and Schwarzer *et al.* (1972) employed different energy intervals for the same purposes.

When spatial resolution is important, surveys are conducted by mounting detectors and equipment on helicopters for low-altitude (15 to 50 m) surveys at slow speeds (100 to 150 km h^{-1}). Normal survey altitudes for fixed wing aircraft are 100 to 150 meters, although Kogan *et al.* (1969) report that 25 to 50 meters were used for snow surveys.

4.6.3 *Background Radiations*

Contributions from cosmic rays, airborne radionuclides, and sources within the aircraft and detector at a typical survey altitude (150 m) can easily account for more than half of the total detector response (Burson *et al.*, 1972).

The cosmic-ray response of NaI(Tl) in aircraft is expected to follow nearly the same altitude dependence as the cosmic-ray ionization rate profile shown in Figure 2-2. Small differences are observed due to latitude, solar cycle variations, aircraft size, and day-to-night differences.

Burson *et al.* (1972) evaluated the cosmic-ray effect from many measurements over inland lakes and the ocean by comparing a selected energy band response to that between 3 and 6 MeV, this latter response being due entirely to cosmic radiation. The cosmic-ray pulse-height distribution was then estimated by subtracting the distributions from two different altitudes, on the assumption that the aircraft, detector, and airborne radioactivity contributions remain constant or are negligible.

The largest contribution and that most difficult to determine is that from airborne radioactivity. Except for airborne radioactivity from local manmade sources, the primary contributors are the radon daughters ^{214}Bi and ^{214}Pb, their concentrations depending on air pressure, wind speed and direction, temperature, soil moisture, and ground cover (Evans, 1969; Moses *et al.*, 1960). Since radon and its daughter concentrations in air are more highly dependent on atmospheric stability conditions and wind patterns than on radon diffusion rates from the ground near the immediate survey area, it is difficult to predict their concentrations, although relative changes over large areas (thousands of square kilometers) can be estimated from pressure and soil moisture conditions. Air filter measurements taken over Lake Mead, Nevada, showed airborne radioactivity concentrations varying by a factor of 20 between successive flights and a strong dependence on air pressure at low altitudes (Burson *et al.*, 1972), high concentrations occurring immediately after or during increased pressure. In less arid areas, soil moisture changes strongly influence radon daughter concentrations (Cox *et al.*, 1970).

The vertical distribution of radon daughters in air must be considered in determining their contribution. Although radioactive equilibrium does not always occur at survey altitudes, non-equilibrium may be negligible during turbulence conditions (at 10 to 100 m altitude as suggested by Staley, 1966). In the absence of inversion layers, which tend to concentrate radon daughters, airborne concentrations were found to decrease slowly and exponentially with altitude (Shopauskac, 1964; Jacobi and André, 1963; Bradley and Pearson, 1970; Newstein *et al.*, 1971).

O'Dell (1972) calculated the gamma-ray response of a NaI(Tl) crystal array to ^{214}Bi and ^{214}Pb distributed in extended layers 15 m thick as a function of height assuming both uniform or exponential radon daughter distribution in the atmosphere. The results in Figure 4-6 show that the detector response is reduced below about 150 m. Experimental points from 10 surveys over Lake Mead tend to verify the calculations (Burson *et al.*, 1972).

Since gamma rays from airborne nuclides are generally isotropically incident on airborne detectors and those from terrestrial radioactivity are not (Beck, 1972b), it may be possible to distinguish between the two by employing separate detector systems. This idea was used for snow surveys in the Soviet Union (Kogan *et al.*, 1969) where an unshielded detector and one with a collimator having an apex angle of 140–160° pointed downward were employed. Purvis and Buckmeier (1969) and Foote (1968) relied on similar dual-detector approaches. These massive shields aid in separating airborne radioactivity components, but require valuable space and weight allotments.

A two-altitude method for accounting for airborne radioactivity is similar to the dual detector method. It assumes that airborne radioactivity contributions are the same for two altitudes and the terrestrial contributions are known. Although this approach is useful in re-analyzing past results or for research purposes, it is not practical for routine operations because extra flights are required.

The gamma-ray spectrum from airborne ^{214}Bi at survey altitudes differs from that from ^{214}Bi in soil, the altitude spectra being "harder" (Fritzsche and Burson, 1973). It may be possible to monitor a wide energy range (such as from 50 to 500 keV and from 500 keV to 3.0 MeV) and infer the airborne component by considering detector sensitivity and contributions from ^{40}K, uranium, thorium, and ^{137}Cs sources in the soil.

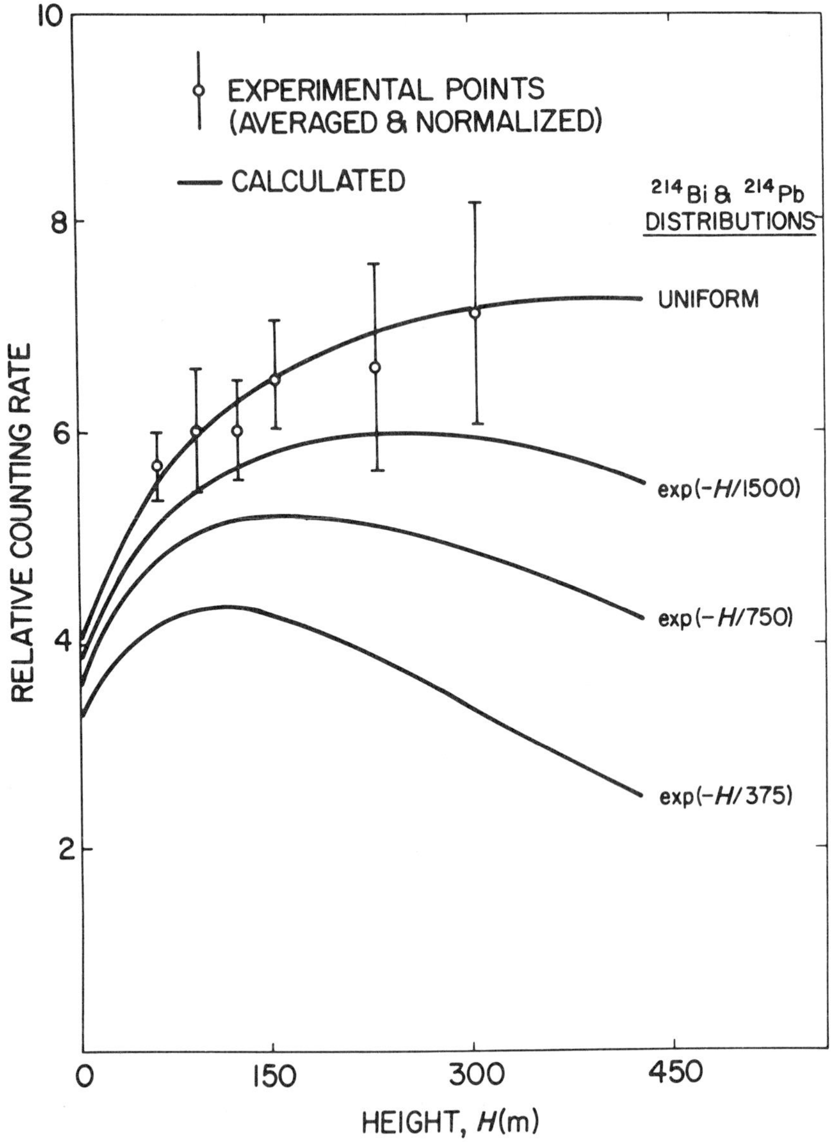

Fig. 4-6. Relative total counting rate in ARMS NaI(Tl) detector for uniform and exponential radon daughter distributions in the atmosphere (from Burson *et al.*, 1972).

Burson and Fritzsche (1972) have suggested several helpful guidelines for dual airborne gamma-ray detectors:

(a) The energy of the radiation measured should be greater than 500 keV, provided that counting statistics can be maintained within 2 percent.

(b) For equal shielding, collimating upward rather than downwards is more effective for discriminating the response to airborne radioactivity from that due to sources on the ground.

(c) Shielding both detector systems in a sandwich arrangement with some of the shielding between the detectors is a good configuration if the placement of shields and detectors maximizes directional responses.

(d) Background contributions should be known to within ±5 percent.

(e) Knowledge of the energy and angular distribution of the airborne and ground radiations is essential.

4.6.4 *Exposure Rate and Radionuclide Concentration Measurements*

To determine the exposure rate one meter above the ground or the radionuclide concentration in the soil, one must characterize the radiation field at survey altitude. Because the photon flux density from natural radioactivity in the soil is low, it is impractical to measure energy and angular distributions, so theoretical calculations are employed (Beck, 1972b; O'Dell, 1971; 1972; Kogan *et al.*, 1969). Calculations indicate that 70 percent of the total gamma flux density at an altitude of 100 m is typically below 250 keV for natural nuclides uniformly distributed in the soil. These photons are isotropically incident on the detector, while about 60 percent of the photons above 500 keV are incident from angles well below the horizontal. For ^{137}Cs exponentially mixed in the soil, virtually all of the primary gamma rays arrive within a cone of ±70° from the normal.

Factors to convert the photon flux density at a survey altitude to the exposure rate at 1 meter above the ground for the natural emitters and ^{137}Cs are shown in Figure 4-7 (Beck, 1972b). Contributions from the different sources vary slowly with altitude and one can use a single conversion factor for natural terrestrial radiation. Errors due to the dominance of a given nuclide would be less than ±10 percent. Correlation of airborne and ground-level measurements above 50 keV shows that 400 counts per second at about 150 meters equals 1 μR h^{-1} one meter above the ground for the fourteen 10.2 × 10.2 cm NaI(Tl) detector array (Burson *et al.*, 1972). The decrease in net count rate with air thickness can be represented by a simple exponential having an exponent of 0.054 cm^{2}g^{-1}. It should be noted that airborne survey results are average ground level values and isolated anomalies of a few meters in dimension may be missed.

Determining ground concentrations of individual nuclides from

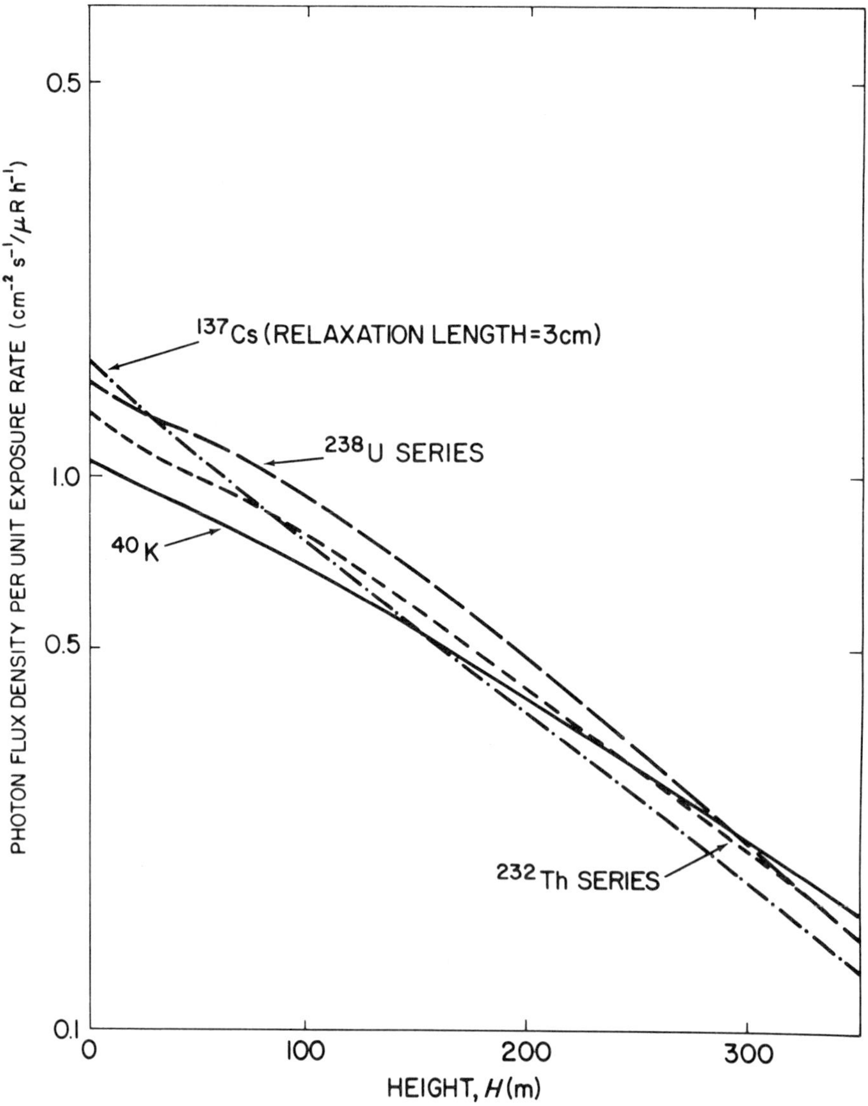

Fig. 4-7. Ratio of photon flux density above 50 keV at height, H, to exposure rate at one meter above ground for uniformly distributed ^{40}K, ^{238}U and ^{232}Th and exponentially distributed ^{137}Cs in ground (from Beck, 1972b).

spectral photopeak measurements is done essentially as indicated in Section 4.5.3. One must determine conversions from photopeak measurements to soil concentrations, taking account of the effect of Compton-scattered photons. These ratios must be determined for each detector system employed.

It is difficult to find locations where the average soil concentrations are adequately known. Grasty and Darnley (1971) employed specially constructed sites and developed procedures for obtaining the conversions. Five separate pads, 7.6 × 7.6 m by 46 cm thick, were fabricated to contain varying, known concentrations of potassium, uranium, and thorium. Measurements were made by parking their aircraft over these pads.

The effective ground area sampled from aircraft is large, about 500 meters in diameter at an altitude of 150 meters for a single measurement. However, the effective area ("ellipse") is elongated because of the aircraft speed and counting times. The average results are influenced by topographic features within the ellipse, such as hills, rivers and ravines, strip mining and strip farming locations, forested areas, wide roads, or residential areas. Hence, one should not compare the average of a few ground-level measurements with that of the airborne survey ellipse. As indicated previously, ground level *in situ* measurements also sample sizeable areas. Lövborg *et al.* (1969) described the volumes of radioactivity-bearing rock as effective "dishes" that detectors in effect sample. Dishes at ground level are fairly deep and are very shallow at aircraft altitudes.

4.6.5 *Applications*

4.6.5.1 *Nuclear Facilities Monitoring.* Aerial surveys of nuclear facilities provide baseline information on the pre-existing terrestrial gamma radiation environment (Doyle, 1972). In the United States over a hundred surveys have been made, covering nearly 80,000 square kilometers. The inferred exposure rates and the primary contributing nuclides provide a basis for the assessment of a radioactivity release or long-term buildup.

The photon counting rate above some discriminator level is recorded frequently along the aircraft path. After background is subtracted, the count rates are adjusted to an equivalent air mass of 16 g cm^{-2} (about 150 meters) and converted to the corresponding exposure rates one meter above ground. Plots are made on U.S. Geological Survey maps and interpolation between flight lines is necessary to construct exposure rate contours.

The correlation of airborne measurements with ground-level data indicates that the average exposure rates in a strip of land 400 meters wide under the flight lines are accurate to within ±20 percent (Burson *et al.*, 1972). Comparison with subsequent surveys is made to detect significant differences. For example Barasch and Beers (1971) have detected buildup of radionuclides along the watershed leaving reactor fuel reprocessing plant sediment ponds and have provided guidance to changes in plant operation.

4.6.5.2 *Determining Water Equivalent of Snow Cover*. Increasing interest in water resources has stimulated searches for better measurements of the water equivalent of snow cover and subsequent runoff correlations (Kogan *et al.*, 1969; Dahl and Ødegaard, 1970; Grasty, 1972; and Peck and Bissell, 1971).

The conventional determination of the water equivalent of snow relies on a network of sample collection measurements. Large errors result, because of the non-uniform distribution of snow and the difficulty of accounting for liquid water or ice in or under the snow pack (Peck, 1972).

Kogan *et al.* (1969) showed that if measurements are made of two monoenergetic gamma rays from the same nuclide (^{214}Bi for example), one can determine the water equivalent from airborne measurements. One measures primary gamma rays from two nuclides if their ratio of concentrations is known, or performs measurements at two different altitudes if their concentrations are not known.

Kogan *et al.* (1969) indicated that attenuation curves for practically any combination of natural radionuclides are nearly independent of the U/Th/K ratios in the ground, so the attenuation of total count rate can be correlated to the water equivalent. Experimental data show that a simple negative exponential adequately represents the total gamma-ray transmission (Burson and Fritzche, 1972).

One can also compare the count rates in particular photopeaks. Dmitriev *et al.* (1970a) indicate the possibility of using the count rate from ^{137}Cs from worldwide fallout deposition, such as in swampy areas where the ^{137}Cs level may be strong compared to natural emitters. Fritzsche *et al.* (1971) showed that measurements of the ^{208}Tl and ^{40}K photopeaks adequately fit a single negative exponential and Burson and Fritzsche (1972) reported 0.076 and 0.064 cm^2 g^{-1} (water equivalent thickness) for the exponential attenuation coefficients for ^{40}K and ^{208}Tl, respectively, values in agreement with those of Dmitriev *et al.* (1970b) for a collimated, wide angle detector. This method can determine the average water equivalent over a survey line a few kilometers in length within about ± 1 cm.

Environmental parameters that affect measurement accuracy are soil moisture changes, plant foliage, redistribution of activity, fertilizer additions, and local variations of snow depths. Kogan *et al.* (1969) report that the radionuclides in plant foliage are less important than those in the soil. Essentially, no radon or thorium daughters exist in plants because of almost complete exhalation. Concentrations of natural uranium, thorium, and potassium in forest trees are much less than in soil, so foliage is essentially a shield rather than an additional source. The biomass of common forests decreases the exposure rates above them by a factor of 1.3 to 1.5 (Kogan *et al.*, 1969).

Plowing tends to redistribute the soil moisture and radioactivity uniformly, since 80 percent of the gamma rays emanating from typical soil come from natural emitters in the top 15 cm (Beck, 1972b). Kogan *et al.* (1969) and Fritzsche *et al.* (1971) concluded that errors in airborne measurements due to normal variations in the snowpack would be less than other effects (about 5 percent for a water equivalent of 10 cm). With the combined parameters contributing to errors in the average water equivalent, the overall accuracy is about ± 0.5 to 1.0 cm, which is sufficient for practical purposes.

Airborne measurements aid in estimating the average water equivalent of the snowpack over a broad area and in flood predictions. Kogan *et al.* (1969) illustrate their value by reporting that 1 to 2 million km^2 have been surveyed in the Soviet Union to develop water equivalent snow cover contour maps for compiling hydrological forecasts.

4.6.5.3 *Soil Moisture Monitoring*. Under idealized conditions (uniform soil moisture and radioactivity concentration), the gamma radiation levels above the ground vary inversely with soil density. Local soil moisture variations and non-uniform nuclide distributions due to plant growth, tillage, soil and rock distributions as well as radon and its daughters affect this relationship, and measurements represent an average condition. Work by Zotimov (1968) and Peck and Bissell (1971) have indicated a few percent reductions in the ^{40}K and ^{208}Tl counts, respectively, for a 10 percent soil moisture increase. Burson and Fritzsche (1972) indicated that the counts in the ^{40}K peak were reduced proportionately as soil density increased, but the influence on the total (gross) count rate was less than predicted for ideal conditions, because of non-uniform distribution of radon daughters in the top few centimeters of soil. Though it is possible to determine average soil moisture changes by observing changes in the gamma flux density above the ground, airborne survey results can be quantitatively interpreted only if some knowledge of soil moisture changes exists. Soil moisture changes of 20 percent are common and produce up to 20 percent changes in the gamma-ray measurements, so they cannot be neglected for accurate water equivalent determinations. McHenry and Gill (1970) report the possibility of using this technique on the ground.

Airborne measurements near Luverne, Minnesota, were made under bare ground conditions and compared with soil samples analyzed for moisture content. Burson and Fritzsche (1972) have indicated that increases in soil moisture decreased the counting rate due to ^{40}K, as expected. Changes in the observed counting rates reflect the effects of fresh plowing, plant foliage, and soil moisture variations with depth. Large variations in the observed soil moisture conditions (±25 percent, standard deviation) also influenced the results. Measurements near Auburn, New York indicated that the average soil moistures compared well to those obtained from ground samples even though soil moisture changes were small. Since soil moisture varies considerably from point to point, obtaining a "true" average soil moisture over a large area requires many soil samples. This shows why airborne surveys have good potential for observing soil moisture changes throughout large areas.

4.6.5.4 *Mineral Exploration.* Airborne surveys of terrestrial gamma rays for the purpose of mineral exploration have been described by, e.g., Darnley and Grasty (1971), Foote (1968), and Schwarzer *et al.* (1972). The gamma ray "signature" of rock and soil types is obtained from photopeak measurements of the principal natural emitters.

To obtain the net responses to nuclides in the soil, consideration must be given to three aspects of data reduction. The cosmic-ray and aircraft backgrounds under all the photopeaks, the airborne contribution of ^{214}Bi, and the count rate from Compton-scattered photons from higher energy gamma rays that end up in the photopeaks of the nuclide being counted must all be subtracted. Identification of specific signatures is then done by evaluating the counting-rate ratios of the different photopeaks. A large increase in the uranium/thorium concentration ratio in the soil or rock indicates the possibility of surface uranium deposits in the area. While gamma-ray surveys for mineral exploration have been conducted for many years, identifying and correlating specific geologic environments by their gamma ray signatures has only just begun.

4.6.5.5 *Other Applications.* Airborne measurements are convenient for locating lost radiation sources and to assist in the rapid assessment of abnormal releases from nuclear facilities.

In 1968 a 330 mCi ^{60}Co source was lost between Salt Lake City and Kansas City, a distance of 1900 kilometers. An aircraft traced the known path of the truck carrying the source at altitudes between 90 and 300 meters above the highway and easily located the source near St. Joseph, Missouri (Hand and Weissman, 1968). A unique search was made for an Air Force test missile carrying two 470 mCi ^{57}Co sources that impacted many kilometers from the White Sands Missile Range. By use of a NaI(Tl) detector array set for an energy band centered on 122 keV, the impact area was located in less than two days (Deal *et al.*, 1971) after ground surveys were fruitless.

Interest in low radiation levels and environmental radiation doses to populations (Oakley, 1972; NAS-NRC, 1972; NCRP, 1975) has emphasized the value of airborne radiation measurements. More than 10 percent of the United States has been surveyed (Doyle, 1972) and a very large portion of the Soviet Union (Kogan *et al.*, 1969). This information can be collected during surveys done for other purposes at little extra cost.

Aerial surveys for plutonium using multiple arrays of NaI(Tl) crystals mounted in helicopters have also shown promise (Stuart, 1971). The technique depends on measurements in an energy band centered at 60 keV, due to ^{241}Am associated with plutonium (Section 4.5.4). If background interference is accounted for by monitoring two adjacent energy bands, ^{241}Am surface concentrations as low as 600 nCi m^{-2} can be detected, but the inference of plutonium concentrations from ^{241}Am measurements is still problematical, as it depends strongly on the radionuclide composition.

Airborne surveys may be the only practical means of monitoring. Stuart *et al.* (1973), for example, described surveys made of the entire Eniwetok Atoll using 40 NaI(Tl) detectors mounted in helicopters. The results included contours of ground level exposure rates and estimated ^{60}Co, ^{137}Cs, and ^{241}Am ground concentrations, as well as natural radiation levels for the Atoll.

4.7 Alpha and Beta Detectors

4.7.1 *Problems of Alpha and Beta Detectors*

Beta rays, primarily from ^{40}K and the uranium series, can interfere with the measurements of the penetrating components and are not negligible contributors to the total ionization or absorbed dose rate in air. Typically, as Table 2-10 shows, the contribution to total ionization is almost half that from natural and present fallout gamma radiation and is about equal to the ionization near sea level from cosmic radiation (Gibson *et al.*, 1969; Iida and Kawano, 1972).

Beta-ray ionization varies considerably with time and depends strongly on the exact location relative to the ground, varying by about a factor of ten between the ground surface and a height of 5 m. The ionization also depends on the type of ground and its composition, especially water content, and the amount of material such as snow on top of the ground. The contribution to air ionization from fallout emitters deposited on the ground following nuclear weapons tests in the atmosphere was about 20 to 40 times natural levels during the mid-1960s, but has decreased to levels equal to those determined for the early 1950s (Gibson *et al.*, 1969; Ikebe, 1970).

In one of the few studies of environmental beta-ray ionization, Iido and Kawano (1972) identified only six literature references between 1951 and 1969. They made differential measurements with a 31-liter thin-window ionization chamber equipped with a 1.1-g cm^{-2} absorber. The beta-ray ionization profile between the ground surface and 1.6 meters was compared with a calculated profile with fair agreement. Differences were attributed to difficulties in distinguishing airborne beta radioactivity and that from the ground. These very difficult measurements will probably continue to have scientific interest. More information on the natural beta-ray field is desirable if direct *in situ* monitoring of manmade beta emitters is contemplated.

The determination of the beta-ray flux density or ionization from manmade emitters requires correction for contributions from the naturally-occurring sources. This is quite difficult because of the intensity, the large vertical gradient (from ground sources), and the strong time variations of the natural beta-ray field (see Section 2.3.4). Except in circumstances of high radionuclide deposition, comparable to the fallout levels existing in 1961–63 or high local concentrations of an atmospheric radionuclide such as ^{85}Kr, determination of manmade beta radionuclides in the environment by direct measurements requires a detailed study of the natural beta plus gamma background at the location of measurement, independent knowledge of the air and ground concentrations of natural beta emitters, and several meteorological parameters. No investigations have yet been carried out in sufficient detail to indicate the practicality of an *in situ* technique for levels from manmade sources comparable to or less than the natural background level.

One important long-term addition to the environment, notably due to increasing reliance on the nuclear fuel cycle, may be ^{85}Kr (Knox and Peterson, 1972). A projection based on this expected growth showed that the 1972 air concentration was about 16 pCi m^{-3} (Jaquish and Johns, 1972), an increase of about a factor of ten during the previous ten years, and the concentration in the year 2000 may be about 1 nCi m^{-3} (Kirk, 1972). These concentrations are a tiny fraction of the limit recommended for the general population (0.3 μCi m^{-3}), but it is prudent to determine concentrations near reactor fuel reprocessing plants, as well as trends in the general environment.

The currently accepted methods of measurement of ^{85}Kr in environmental air involve sample collection and subsequent treatment and counting in a laboratory (Section 6). *In situ* ^{85}Kr measurement methods have been attempted, notably by Smith *et al.* (1970), but sensitivity levels appear to be well above those required for measuring the present ambient concentrations. If environmental monitoring is contemplated, thin-wall detectors, perhaps rugged G.M. counters, would be useful if properly field tested. Smith *et al.* (1970) indicated that the most sensitive commercial G.M.-detector system was too fragile for reliable field use, but it served as guidance for the design of the more rugged G.M. counter array described by Beck and Freeswick (1975). The beta-ray dose corresponding to measurable short-term increases near a reactor fuel reprocessing facility may account for most of the total ^{85}Kr dose one could attribute to plant operation.

Measurements of naturally-occurring radon and thoron and their daughters in air are made either to determine the dose or to account for interference in the field measurement of other alpha- and beta-emitting nuclides.

The dose to the bronchial epithelium from environmental radionuclides is delivered primarily by the alpha-emitting radon daughters, ^{214}Po and ^{218}Po. Typical background levels result in estimated absorbed dose rates that range between 2 mrad y^{-1} and 1 rad y^{-1} (UNSCEAR, 1972). The concentrations of radon and thoron and their daughters fluctuate markedly with time, the outdoor fluctuations being controlled by meteorological variables such as barometric pressure and windspeed and the indoor levels by the types of building structure and ventilation, and by living habits (Hultqvist, 1956).

The measurement of environmental radon and thoron is usually accomplished by collecting their daughters on a filter and measuring the subsequent alpha and beta decay as a function of time. Analysis of sequential alpha counting data as a function of time allows evaluation of the concentrations of radon, thoron, and their short-lived daughters (Raabe and Wrenn, 1974). Alternatively, samples of air may be collected and the alpha-emitting radon or thoron and daughters measured by pulse counting techniques (Hultqvist, 1956; Lucas, 1957). Various techniques for measuring radon and thoron and their daughter concentrations in air have recently been reviewed by Budnitz (1973).

Recently Spitz and Wrenn (1974) described a pulse counting scintillation instrument capable of continuously measuring environmental radon above concentrations of 100 pCi m^{-3}. This instrument is useful for determining daily, weekly, and seasonal fluctuations of radon indoors and, with increased sensitivity, would be useful for lower outdoor concentrations.

4.7.2 *Possible Improvements for Alpha Detection*

The transuranic nuclides present very difficult measurement problems, which ought to be studied in anticipation of the greater quantities of these radionuclides coming into use and possibly being dispersed into the environment. Present *in situ* photon measurement techniques for ^{239}Pu and ^{241}Am, discussed in Section 4.5.4, are insufficiently sensitive and reliable for determining small concentrations. Their primary value is for determining localized, relatively large concentrations and for providing guidance to the collection of samples.

Investigations are beginning which may improve the sensitivity and reliability of the *in situ* determination of ^{239}Pu and ^{241}Am possibly by a factor of ten by employing high resolution Ge(Li) spectrometry suitable for low-energy photon spectrometry (Armantrout *et al.*, 1974; Roth and Huckabay, 1974). Direct *in situ* alpha counting measurements do not appear practical (Healy, 1971), although techniques based on such measurements are worthy of further investigation.

5. Collection and Preparation of Samples for Laboratory Analysis

5.1 Introduction

5.1.1 *Sample Collection Considerations*

Determinations of radionuclide concentrations in environmental materials collected and brought into a low-level radiochemical laboratory for analysis are the major part of any environmental radioactivity measurements program.

In some circumstances *in situ* measurements are sufficient for dose assessment, but often they are combined with sampling and laboratory analysis. The latter determinations usually yield better descriptions of the analyzed material, greater analytical accuracy, and the required sensitivity for radionuclides that do not emit penetrating radiation. On the other hand, sampling requires greater efforts, provides information less promptly, and requires additional interpretation for relating analytical results to the required environmental information.

Samples are either taken manually from the environment or accumulated in unattended devices and analyzed in the laboratory or possibly in the field. Some collection devices have detector and recording systems that measure sample emissions during collection. Most environmental media have been sampled at one time or another, but often somewhat indiscriminately. Sampling should be selective where possible; for example, air may be collected in its entirety or passed through a collector that retains only one constituent such as the particulates, water vapor, or iodine.

There is not extensive published guidance on sample collection principles beyond that referred to in Section 3.1.2. Substantial information is available on sample collection equipment and methods with appropriate equipment as described in Section 5.2.2. The design of an effective monitoring program, therefore, requires access to a substantial amount of environmental data and often extensive sampling.

Once the most significant pathways from source to man are identified, e.g., Section 3.1, further monitoring can depend on analyses of fewer samples for relatively few nuclides. The relative doses from specific radionuclides may not be in proportion to their concentrations in the air, water, food, or other material that are in the pathways for introduction of the radionuclides into the body. Thus, a particular radionuclide in a mixture may be the most important one, although its measurement is complicated by its low relative concentration. The dosimetric importance of radionuclides depends on their differential metabolism by man, the quantity consumed, and the physical and chemical properties that determine their movement in the environment and eventual uptake, deposition, and retention by man. As a result, the measurements should be specific for given radionuclides and in some cases should identify their chemical and physical forms.

5.1.2 *Sample Analysis Considerations*

It is desirable to select analytical techniques requiring little chemical separation, and to rely on the measurement method for developing specific nuclide information. This generally results in smaller analytical errors and costs. If the desired specificity is obtained in the sampling process or by chemical separation, the specificity requirements of the laboratory measurement system are minimized.

In the early days of nuclear facilities monitoring, samples were analyzed to determine their so-called "total alpha, beta, or gamma" activities, which were then compared for complicance with recommended values for unknown mixtures. Such total activity measurements continue in use, although they have little value for dose determinations or scientific studies (ICRP, 1965a). Total activity measurements may provide an inexpensive and approximate indication of totals and changes near a nuclear facility or be useful in a low-level laboratory as a method for classifying samples. In samples that have relatively small or constant amounts of radioactivity from background radionuclides, the measurement of total activity can assure that important radionuclides were not missed in measurements of specific radionuclides. Such measurements are a possible operational tool, but should not be used for documenting levels of environmental radioactivity (see Section 6.2.4).

This section is concerned with the principles and techniques for collecting, handling, and processing environmental samples. The methods described concern principally samples having very low to moderate radionuclide concentrations but are often applicable with simple modifications to samples having higher concentrations. A low-level concentration is that level at which the desired precision of counting is difficult to achieve in a reasonable counting time and/or that level at which the counting rate from the radioactive sample is equal to or less than the background counting rate.

Low-level radiochemical separations and measurements have been treated in many books, monographs, and reports (see, for example, ICRU, 1972; Kahn, 1972; Watt and Ramsden, 1963, Sugihara, 1961; Crouthamel and Heinrich, 1971). Although not specifically concerned with radioactivity, the monograph by Korenman (1966) discusses many of the factors and considerations that apply to all analyses at low concentrations, including separation methods and problems in handling solutions containing very low concentrations.

Most of the principles, methods, and techniques used in the measurement of radioactive constituents in environmental samples at low levels are similar to those applicable to high-level measurements. However, in chemical separations extra precautions are required to obtain a low blank value and the chemical procedures may be greatly complicated by the need to process large amounts of complex and intractable substances. Similarly, complicated modifications of basic detectors may be needed to reduce the background and retain high efficiency. This is in contrast to the chemical analysis of stable elements, where the basic gravimetric and volumetric methods used to measure ordinary quantities (milligrams) cannot be used to determine traces (nanograms) of the same elements, so that a different analytical method, e.g., atomic absorption spectrometry, or enzymatic reactions, must be used.

Low-level radioactivity analyses should be conducted in a laboratory set aside for that purpose. If even moderate quantities of radioactive materials are used in a low-level facility, contamination of the low-level samples, counters, and chemical reagents may occur. Since the desired analyses are complex and considerable expertise and instrumentation are required, these analyses should be done by special low-level analytical groups rather than by the staffs of individual plants.

The successful measurement of low activities requires an optimum combination of high detection efficiency and low (and equally important) stable detector backgrounds. The sensitivity of measurements varies (approximately) directly with detector efficiency and inversely with the square root of background. For this reason, the more sensitive of two detectors may not exhibit the lower background. Accuracy and precision of low-level measurements depend strongly on the background stability, and since long measurement times are required for determining low concentrations, a somewhat higher but more stable background, particularly if accompanied by higher efficiency, may be preferable to a lower but more variable background. Increased sensitivity can also be obtained by increasing the sample size. To avoid a proportionate decrease in detection efficiency, very large environmental sample sizes must sometimes be processed to concentrate the activity. This in turn introduces the problems of obtaining adequate chemical recovery and radiochemical purity of the desired radionuclides, determining the proper reagent blank, and avoiding sample contamination (Sugihara, 1961). Naturally-occurring and synthetically produced radionuclides in low abundance are widespread, and the radioactive content of chemical reagents, materials of construction for radiation detectors and shields, laboratory and sampling supplies (such as filter paper), and laboratory air (when precipitates are filtered) must be carefully considered and evaluated.

Contamination of materials and reagents is discussed in two ICRU reports (ICRU, 1963; 1972). Large variations in the alpha-, beta- and gamma-ray activities in various metal samples have been reported by DeVoe (1962) and Weller (1964) and Rodriguez-Pasqués *et al.* (1972). For example, the beta radioactivity in 12 different lead samples differed by a factor of about 20, as determined in a 4π windowless counter.

A significant effort should be devoted to quality control and assurance. This effort should include the regular analysis of standard and duplicate samples, including unidentified duplicates, frequent blank determinations, regular tests of counters with calibrated sources, and careful examination of the routine results and counter performance. Interlaboratory analyses and calibration programs are important since they provide independent validity tests. Interlaboratory analytical programs for environmental materials are conducted by the U.S. Environmental Protection Agency's Environmental Monitoring and Support Laboratory in Las Vegas, Nevada and the International Atomic Energy Agency, at its laboratories in Vienna and Monaco. The National Bureau of Standards provides certified standards for some radionuclides and some standard reference environmental materials. Commercial sources are available for many radionuclides not standardized routinely by the Bureau, and frequently these standards have been previously compared with Bureau standards. The user should satisfy himself by whatever means are at his disposal as to the quality of the standardized sources he uses.

The amount of time to be spent on quality control will vary with the particular analytical procedure and the experience of the laboratory. At least 5 to 10 percent of the analyses and measurements should be devoted to blanks, standards, and duplicates. It has been recommended that approximately 15 percent of the samples in a given laboratory be analyzed as unidentified duplicates (WHO, 1968) and that 10 percent of all analyses should be devoted to quality assurance (EPA, 1972).

Not all environmental samples contain "low-level" concentrations of radionuclides. For example, air, airborne particulate, and precipitation samples collected near and shortly after nuclear test detonations or some water samples near nuclear facilities may contain sufficient radioactivity for conventional measurements. Concentrations of natural activity also occur at moderate levels. The ^{222}Rn content of air over land masses is in the range of 10 to 1000 pCi m^{-3} (see Figure 2-1) and at this concentration, several cubic meters of air are sufficient for measurement by standard counting techniques. Similarly, the natural ^{40}K content of clay soil ($\sim$15 pCi g^{-1}) and many natural waters (5 to 10 pCi l^{-1}) is relatively easy to detect by ordinary beta counting.

In this section, methods for collecting and preparing the sample for measurement are discussed, including chemical separation if needed. The final measurements are generally by radioactivity detection, except that determinations of long-lived, low specific-activity nuclides, such as ^{238}U and ^{232}Th are based on fluorimetric techniques, because they are more sensitive or more accurate. For medium half-life heavy nuclides, such as ^{239}Pu and ^{241}Am, special mass spectrometry is more sensitive, although careful radioactivity measurement is adequate.

A number of laboratories engaged in the radiochemical analysis of environmental samples and other organizations have published manuals that contain suitable analytical procedures (see, for example, Douglas, 1967; Harley, 1972; Johns, 1970; Krieger *et al.*, 1966; Guthrie and Grummitt, 1963; Porter *et al.*, 1965; Krieger and Gold, 1973; Sheehy, 1965; WHO, 1966). In the use of these manuals, one should evaluate procedures developed before about 1966 for obsolescence in whole or in part, because of advances in analytical radiochemistry in recent years, especially in detection systems. In particular, advances in gamma-ray spectrometry with germanium semiconductor detectors, in alpha particle spectrometry with silicon semiconductor detectors, and in liquid scintillation counting and spectrometry for alpha and beta emitters make these detection techniques the methods of choice. Although improved ion-exchange and liquid-liquid extraction separation techniques are also available, there is little reason to perform chemical separations and alpha or beta counting for radionuclides with a reasonable gamma-ray abundance in their decay schemes if one has access to a system having adequate energy resolution, such as a Ge(Li) detector coupled to a pulse-height analyzer.

References in the literature, which are informative about analytical radiochemistry, include Broda and Schonfeld (1966), Choppin (1961), Friedlander *et al.* (1964), Haissinsky (1964), McKay (1971), Overman and Clark (1960), and Wahl and Bonner (1951). Although some do not specifically concern environmental sample preparation and analysis, many of the principles, separations, and detection methods contained in them are applicable.

5.2 Types of Environmental Sampling

5.2.1 *Problems in Sampling*

There are three main considerations in sampling radionuclides beyond those shared with other types of environmental sampling. With active sample collectors, such as filters and absorption columns, the extent of retention will depend on the physical-chemical forms of the radionuclide (of which there may be several, including unexpected ones at the expected low levels) and on interfering substances. Secondly, in the sampling of some material, such as vegetation and soil, uneven distribution of radionuclides is common, but difficult to observe at the encountered low levels. Thirdly, the characteristics of the emitted radiation must be considered in devising the sampling program.

The effectiveness of collection of a single form of the radionuclide on a filter or column is tested by passing the air or water to be sampled through several of these collectors in series. If each identical collector has the same removal efficiency, E, then the fraction of the total removed by the nth collector, F_n, is

$$F_n = E(1 - E)^{n-1} \tag{5-1}$$

Thus, E can be determined by measuring the relative amounts removed by successive collectors; for two collectors, for example,

$$E = 1 - \frac{F_2}{F_1} \tag{5-2}$$

The use of more than two identical collectors in series can indicate when several species with different retention efficiencies are present, because the values of E computed for successive pairs of collectors will then differ. A "train" of collectors, in which the several sets of component collectors are each specific for one form of the radionuclide, is needed if more than a single form must be collected. The distribution of the radionuclide among various forms, however, can change with time during collection, and may actually be affected by the collection process.

Uneven distribution of a radionuclide in environmental materials depends on the pathway (e.g., surface deposition as compared to root uptake in vegetation), as well as inhomogeneity of the medium. An obvious example of the latter is the variation in radionuclide concentrations among animal organs. Radionuclide distribution in a medium, hence, can and does often differ from that of its stable form as found in nature. Some understanding of this distribution must be achieved before routine sampling is begun so that samples can be adequately selected and defined with regard to the purpose of the program.

An important property of a radionuclide to be considered is its half life. Measurements have to be corrected for radioactive decay between sampling and analysis and more complex corrections are needed for radionuclides comprising decay chains. Equally important, reports of no detectable radioactivity in a sample should include information on the time interval between sampling and measurement to indicate the detection limit for short-lived radionuclides. The type of radiation emitted by a radionuclide affects sample selection and eventual treatment because relatively non-penetrating radiation such as alpha particles must be measured in very small amounts of material, while gamma rays can be conveniently detected in bulk or unprocessed material.

A common problem in collecting and storing a sample of gas or liquid is loss of the radionuclide to apparatus and container surfaces (see Section 5.3.4). Selection of relatively nonretentive materials and minimization of areas and collection and sample transit time are recommended. Information on the magnitude and rate of uptake by surfaces is widely scattered and qualitative, however, and no materials appropriate to the entire spectrum of encountered radionuclides are known.

Some guidance on sampling is usually provided with standard methods (APHA, 1971) and is specifically available for seawater sampling (IAEA, 1971c). To minimize losses during sample collection and storage, containers of appropriate composition should be used. The liquid contents can be frozen, acidified, or otherwise modified, by adding a complexing agent or stable "carrier". It may be possible to check the collector or container by leaching it and analyzing the leaching solution, or by directly measuring radiation from its surfaces. Comparisons of results obtained with several collectors or containers will usually indicate that a problem exists. Material balances and observation of concentration ratios for several radionuclides will also provide this warning.

Changes in the physical or chemical form of a radionuclide during collection or storage can occur in all types of materials. If, in addition to requiring accurate total radionuclide concentration, the physical-chemical form of the radionuclide is to be described, a detailed and difficult testing program may be required so that the initial forms can be retained for analysis. Separation of these different forms during collection in the field may prevent changes and aid in yielding the required information.

5.2.2 *Types of Sampling*

The sampling described here is directed primarily toward radiological monitoring in the environs of nuclear facilities for determining extremely low concentrations of radionuclides. Specific collection procedures and apparatus are described in various radioactivity surveillance manuals (Harley, 1972; EPA, 1972; Setter, *et al.*, 1966) and in other collections of standard methods cited in Section 5.1. Special methods used in geophysical and geochemical studies are not described. Sampling is categorized here as atmospheric, terrestrial, and aquatic.

The general principles that apply to all sampling activities include:

(a) *Sources of Information*

Available information concerning the sources and modes of transfer through the environment of radionuclides is needed to guide the selection of types and location of *in situ* measurements or sample collections and analyses, as well as the sample size and collection frequency in order to identify the radionuclides of interest. Particular guidance on sample selection and interpretation should be obtained from specialists in pertinent fields such as agronomy, aquatic biology, meteorology, limnology, animal husbandry, or wildlife management.

(b) *Analytical Considerations*

Guidance from the radiochemical analyst is needed to assure that all necessary parts of the sample are collected, that the sample is adequately preserved, and that loss of radioactivity between the time of the sampling and analysis is known or negligible. The amounts of samples depend on the radionuclide concentration, the sensitivity of the radiation instruments, the capability of analytical methods, and the coordination of sampling and laboratory schedules.

(c) *Significant Data*

Sufficient replicate samples should be collected to determine concentrations and the associated mean errors. Sufficient background samples are also needed to aid in distinguishing pre-existing nuclides from those under consideration.

(d) *Modeling*

Information concerning the physical and chemical characteristics of samples which influence radionuclide retention and turnover is needed for dose assessment. The concentration of stable nuclides is usually important in predicting the retention of chemically similar radionuclides in samples or in man.

5.2.3 *Atmospheric Sampling*

5.2.3.1 *Air*. Air is sampled to evaluate radiation exposures from external or inhaled radionuclides in gaseous or particulate form, e.g., ^{3}H, ^{14}C, ^{131}I. Volumes of 0.1 m^3 or more may be required to accumulate sufficient radioactivity. Low volumes may be collected in plastic bags with a hand pump. Large volumes of air are collected in metal tanks by pumping to pressures of 10 to 30 atmospheres (Karches *et al.*, 1971). Data on atmospheric stability, wind speed, and wind direction under which the samples were collected should be noted. Some differences in radionuclide concentration can be expected from moment to moment even under relatively constant conditions. The naturally-occurring radioactive gases, ^{222}Rn and ^{220}Rn, found at most locations, may interfere with the eventual analyses.

5.2.3.2 *Gaseous Radionuclides*. Selective collection devices are convenient when only one or a few gaseous radionuclides are to be measured. Noble gases and gaseous radioiodine can be sampled with charcoal or "molecular sieve" collectors (Koch and Grandy, 1957; Megaw and May, 1962; Johns, 1973). Water vapor for ^{3}H analyses may be retained on a desiccant, condensed on a cold surface, or collected in water (Jacobs, 1968; NCRP, 1976).

If the concentration is needed, the volume of air is measured (in terms of flow rate and collection time); if specific activity (radionuclide activity per total amount of element) is to be measured, a passive device, such as a bag of silica gel suspended in air to collect tritiated water, may be satisfactory. Retention efficiency must be tested under operating conditions and possible adverse effects on the collection efficiency by other gases at various concentrations should be tested. The chemical form of the radionuclide especially influences the collection efficiency.

5.2.3.3 *Airborne Particles*. Radioactive airborne particles are measured because they cause radiation exposure when inhaled, deposited on surfaces, or ingested. Hundreds to thousands of cubic meters of air can be passed continuously through filters to collect particles. A wide variety of filters (glass fiber, paper, cellulose-asbestos fiber, for example) is available. The type of filter is selected according to needs for high collection efficiency, particle-size selectivity, retention of gaseous iodine with the particles, and ease of radiochemical analysis. It may be desirable to change filter media for specific purposes, or to use several filters in series.

High sampling rates are achieved with high capacity blowers and large area membrane filters (Perkins, 1963; Rieck, 1967). Cellulose-asbestos filter papers combine fairly high efficiency, high flow rates, high mechanical strength, and low pressure drops when loaded and are very useful for collecting large samples, but present difficulties in dissolution. Pure cellulose papers are useful for samples to be dissolved and analyzed radiochemically, but the analytical filter papers used to filter solutions are inefficient collectors for aerosols and clog easily. Of particular value for samples to be analyzed radiochemically are plastic filters such as those composed of polystyrene fibers. They are efficient and capable of sustaining high air flow rates without clogging. They are readily destroyed in analysis, by ignition at low temperatures (~300°C), or by wet ashing with oxidizing agents, and are also soluble in many organic liquids. They have the disadvantage of very low mechanical and tensile strength, and must be handled carefully. Fiberglass filters can function efficiently at high flow rates, but require fluoride treatment for dissolution, and generally contain sufficient radionuclides, e.g., ^{40}K, to complicate low activity analysis.

Evaluations of retention efficiencies have been reported (Lindeken, 1961; Lockhart and Patterson, 1962; 1964; Lockhart *et al*., 1964), but additional tests under operating conditions are desirable. Information on filters is sometimes available from manufacturers. The collection efficiency may change during operation if too many solids accumulate. Electrostatic precipitators, cyclone collectors, and cascade impactors have also been used to collect particles and classify them according to size. Radionuclides on air filters usually include cosmogenic ^{7}Be, decay products of ^{222}Rn and ^{220}Rn, and long-lived fallout from atmospheric nuclear tests.

5.2.3.4 *Deposition*. Dry and wet deposition of radionuclides are measured to follow the movement of radionuclides from air to terrestrial and aquatic media. In principle, a collector with an open surface accumulates dust or precipitation for a specified time, and this material is then analyzed for radionuclide content. The container should be washed to collect radionuclides that tend to adhere to surfaces or lined with a plastic sheet that is analyzed with the container contents. Water may be funneled through a filter or ion-exchange system for collecting some radionuclides. Dry deposition is determined separately on material such as gummed paper or fibers that simulate grass. Precipitation may be collected separately in a device with a cover that opens only during rainfall. To obtain concentration as well as deposition values, the amount of rainfall is measured after each collection. The same radionuclides are usually deposited in rain and collected with airborne particles, but relative amounts depend on the initial vertical distribution, particle sizes, and other characteristics. A collector intended to measure representative deposition should be placed to avoid effects by local factors, e.g., air turbulence, shielding by buildings and vegetation, and sources of dust, fumes, and moisture.

5.2.4 *Terrestrial Sampling*

5.2.4.1 *Vegetation*. Samples of food are of greatest interest because their analysis yields values of the radionuclide intake that contributes directly to dose. Animal feeds, especially forage and field corn, provide important data for determining radionuclide concentrations in the food chain. Analysis of vegetation, such as grass and leaves, indicates the identity and amount of radionuclides in the environment. Foods may be categorized as leafy vegetables, grains, tree-grown fruits, etc., and representative samples selected for analysis. Some materials have particular utility for identification: Spanish moss on trees being a natural filter for airborne particles; ground mosses and lichens integrating radionuclide deposition over many years; and the leaves of perennial plants being useful for identifying nuclides collected in a particular growing season.

The amount of a vegetation sample to be collected may be several kilograms. Smaller samples should be collected whenever the required radionuclide detection limit can be attained, because processing such samples is simpler (Section 5.3.5). Samples must be weighed in a consistent manner so that the concentrations can be compared among samples. Sample mass should be determined at collection, and then air-dried, oven-dried, or ashed.

If samples are collected to derive transfer coefficients or to represent certain exposure pathways, sample selection must be guided by factors such as growth patterns, wind direction frequencies, soil types, and farming practices. The amounts required for such analyses may refer to the entire sample,to its edible part, or to the part that collects or concentrates the radionuclides of interest. One should consider the pathway of radionuclides to the analyzed fractions, such as surface deposition, root uptake, and translocation from other parts of the plant.

As in all terrestrial samples, naturally-occurring ^{40}K and the uranium and thorium series contribute to the background radiation. Deposited fallout, such as naturally-occurring ^{7}Be and the constituents of weapons fallout, may interfere with facilities' monitoring efforts.

5.2.4.2 *Animals*. Animals whose meat is eaten by man are of greatest interest as samples. Other animals may be collected as indicators because they concentrate certain radionuclides, consume foods of special interest, forage over wide areas or inhabit particular locales. Food animals can be collected at the normal time of slaughter; wildlife samples, from hunters, after accidental death (deer hit by autos, for example), or by request to the appropriate state fish and game agency. Wildlife that is relatively rare should not be used in environmental sampling and measurement.

The animal is usually dissected to separate the edible portions of those tissues that are to a degree specific for radionuclides of interest. Unless the animal is very small, tissue from each animal is considered as a separate sample. Where the sample includes considerable water, it must be weighed at uniform water content. If a large amount of sample is available, approximately 2 kg may be taken. Because radionuclide concentrations in animal tissues tend to range widely, a sufficient number of animals must be collected to define the normal range of radionuclide concentrations.

Information and expert guidance should be obtained concerning factors affecting radionuclide uptake and retention such as size, age, sex, feeding locus, and food consumption. The cause and date of death is also significant information. A direct measurement of the radionuclide intake of the animal just before its death is provided by analyzing the stomach content, especially the rumen in cattle and deer.

5.2.4.3 *Animal Products*. The most common type of sample in this category is milk, but other samples, such as cheese and eggs, may be considered. Cow's milk may be a major dietary source of ^{89}Sr, ^{90}Sr, ^{131}I, ^{137}Cs, and ^{140}Ba. Naturally-occurring ^{40}K is in all milk. Milk is one of the few foods that is commonly consumed soon after production, and thus may also contain relatively short-lived radionuclides when consumed. A 4-liter sample of milk is usually analyzed. The milk sample may be taken to be representative of the production of a particular herd and milking time, or aliquots can be combined to represent a longer time period or a larger milk pool. Liquid milk must be preserved for long-term storage. Common chemical preservatives are formaldehyde or thimerosal (sodium ethylmercurothiosalicylate). The effect of the preservative on subsequent chemical separation must be considered. For example, it has been shown that formaldehyde interferes with the exchange of iodine in milk onto synthetic resins because it promotes protein-binding of the inorganic iodine (Kahn, 1965; Murthy and Campbell, 1966; Lamanna *et al.*, 1965). Thiouracil (2-thio-4-oxypyrimidine) has been used to inhibit this protein-binding, and the addition of iodine carrier with the formaldehyde also essentially eliminates this problem, presumably by dilution of the radioiodine with stable iodine (a "holdback carrier"). Low temperatures should be used for preservation if facilities are available. Refrigeration near the freezing point can be used for one to two weeks, and freezing and storage at frozen-food temperatures (about −15°C) is suitable for longer periods. It may be useful to affix temperature-sensitive color tape to the container to indicate the highest temperature the sample has reached during shipment and storage.

5.2.4.4 *Soil.* Soil sample analysis is a common method of environmental monitoring, but the inference of dose to man from measured radionuclide concentrations is complex and uncertain due to wide variations in resuspension or uptake by vegetation. Furthermore, collection, preparation and analysis are tedious and costly. Uptake from soil through the root system usually contributes only a small fraction of the total radionuclides in vegetation during periods of radioactive fallout, but may subsequently be the only pathway. Because soil accumulates long-lived radionuclides, information from soil sample analyses is not very useful for estimating short-term trends in radionuclide concentrations, but can, in some circumstances, be used for retrospective evaluations of environmental contamination.

It is sometimes useful to analyze gamma-ray emitting nuclides in soil as an aid to evaluating direct radiation exposure above ground. The determination of radionuclide concentrations in soil by *in situ* spectrometry (Section 4.5.3) is useful if the vertical radionuclide profile is known. Approximately 1 kg of soil is a suitable practical amount for laboratory gamma-ray spectral analysis. Both the collection area and depth should be measured. Because the concentration of deposited radionuclides usually decreases rapidly with depth in undisturbed soil, one may have to sample as a function of depth (see Figure 4-4). The sample should also be defined according to soil characteristics, agricultural status, and vegetation cover. The latter should be analyzed separately.

5.2.5 *Aquatic Sampling*

5.2.5.1 *Water*. Water samples are collected near discharge points at nuclear facilities to determine compliance with prescribed radionuclide concentration limits and from water supplies to evaluate radiation exposure through ingestion. The volume of water needed to assure compliance is often only a few milliliters, but many liters have to be analyzed to measure actual radionuclide concentrations in locations away from the facility. Water is collected sometimes continuously by proportional samplers, sometimes at selected intervals in response to batch discharges, and occasionally randomly. The major concerns are to assure collection of a representative sample and to maintain radionuclides in their original form. To minimize adsorption on suspended material, this material should be separated by filtering at collection time (see Section 5.2.1).

5.2.5.2 *Dissolved Radionuclides*. Dissolved radionuclides, unlike much of the suspended material, may be transported in water supplies to distant locations and be consumed or used for irrigating crops. Specific dissolved radionuclides of interest are frequently separated from the water sample on site to avoid transporting large volumes of water. Ion-exchange resins, inorganic ion-exchange materials, surface adsorption media, organic solvents, and precipitated salts are useful for this purpose. Ion-exchange resins are particularly applicable to fresh water because numerous radionuclides are often retained by the same resin. In brackish water and seawater, however, the salt content precludes retention except for a few radionuclides or by specially prepared resins. The efficiency of the system or combination of systems must be tested under operating conditions to determine the effects of flow rate, volume, chemical form of the radionuclide, stable isotopes of the radionuclide, and other material in the water.

5.2.5.3 *Radionuclides in Suspended Material*. Radionuclides are measured in undissolved material to evaluate deposition and accumulation in sediments of bodies of water. Samples representing several hundred liters of water can be collected by continuous or batchwise pumping through filters. Paper, membrane, polystyrene, glass fiber, or asbestos filters may be used, depending on requirements of high volume, particle diameter specificity, and ease of chemical analysis. Several filters of different pore size in series provide separation by particle size. Settling and continuous batch centrifugation are also used for separation. Possible adsorption of dissolved radionuclides on the filter and filtered material must be considered.

5.2.5.4 *Fish*. Fish are measured for radionuclide content to quantify dietary intake by humans, and secondarily, as indicators or integrators of radioactivity in the ecosystem. Analysis of food fish, prepared for consumption of its edible portions, is of major interest. The whole fish is analyzed if it is used for the preparation of fish meal for consumption. In a detailed program, fish are collected by species; in less detail, fish typical of categories such as bottom feeders, insect eaters, and predators may be collected. Fish used as indicators may be analyzed whole, or dissected for tissues that concentrate specific radionuclides.

Several kilograms of each fish sample are usually required; this may be one large fish or many small ones. Numerous replicate samples are desirable because of the usual difficulty in knowing whether a fish caught at a location had lived there for an extended period. Thus, the presence or absence of a radionuclide does not permit any definite conclusion concerning the presence of the radionuclide in water at that location. For some fish, more specific information concerning their location may be available; for example, dams, salinity gradients, and temperature gradients can be effective barriers to their movement. Expert guidance and information on fish age, feeding habits, and the quality of the aquatic environment are desirable to evaluate the significance of any findings.

5.2.5.5 *Shellfish*. Like fish, shellfish such as clams or oysters serve as radionuclide indicators and they have the further advantage of being relatively stationary. The shells, meat, and in some cases the fluid, are analyzed separately. However, shellfish being filter feeders, contain much sediment in the intestine. If this material is not also separated, the results can be seriously misleading.

5.2.5.6 *Aquatic Plants*. Aquatic plants are indicators and concentrators of radionuclides, and can be vectors in the water-man food chain, as indicated for several pathways in Section 3.2.3. If available, several kilograms should be collected per sample. Algae are obtained by filtering water or by scraping surfaces submerged in water. Partially submerged plants, rooted beneath water, and macroalgae should be collected by species. For microalgae, it would be difficult to separate species from each other and from other material in sufficient quantity to permit analysis.

5.2.5.7 *Benthic Material*. Sediment is an indicator and accumulator of radionuclides that are insoluble or adsorbed on insoluble material on the aquatic system. The most useful sample is collected by hand or diving, because location and depth are then defined. Conventional collection is by a dredge dropped from a boat. The dredge mechanism is activated by contact with the benthos, hence surface sediment is collected without accurate knowledge of the location or the depth. A 1-kg sample is usually sufficient. For river sediments, it is of interest to describe the stream flow conditions that lead to benthal deposition and the movement of sediment, as well as the characteristics of the sediment such as particle-size distribution, soil type, ion-exchange capacity, and organic content.

Sediments at shorelines and river- or lake-banks are collected to determine direct radiation exposure while bathing or fishing, although *in situ* measurements are often preferred. Collected samples are treated similarly to soil samples (Section 5.2.4) except that the accumulation from and interaction with the aquatic environment must be considered rather than deposition from air and uptake by grass.

5.3 Sample Preparation

5.3.1 *Purposes*

Sample preparation prior to measurement may be required: (a) to place the sample in suitable form for measurement, without separating the radionuclides of interest; (b) to distinguish the desired nuclides from others that interfere with the final measurement by producing a response in the detector (e.g., other radionuclides in the sample or, in liquid scintillation counting, luminescent materials); (c) to increase the specific activity so that the radionuclides can be measured with improved sensitivity or accuracy; or (d) to reduce self-absorption in the sample matrix for measurement of alpha and pure beta emitters and low energy photons.

As few steps as possible should be used in preparing samples for measurement to reduce the possibility of cross-contamination, sample processing errors, and cost.

Sample preparation without recourse to specific chemical separations is limited to total alpha-, beta-, and gamma-ray counting and gamma-ray spectrometry. In these cases, one needs a uniform distribution of the radioactive substances throughout the sample and the same sample size, thickness, and configuration within each group of samples. Analytical laboratories "calibrate" these samples with locally-made standards having the same physical properties and known quantities of the appropriate radionuclides, in addition to relying on standards obtained from the National Bureau of Standards. Sample homogeneity becomes more important as the range of the radiation becomes shorter, but since large samples are generally used for low-level gamma-ray spectrometry, homogeneity is also a significant consideration in this measurement, since self-absorption and the physical sample-detector geometry may vary in different parts of large samples. The homogeneity problem in liquid samples is discussed in Section 5.3.4.

5.3.2 *Preparation of Samples*

5.3.2.1 *Destruction of Organic Matter*. The first step in obtaining a solution for analysis is usually the destruction of organic material in the sample. Materials that contain little organic matter, e.g., some soils, can be dissolved directly, since the dissolution treatment, with oxidizing acids or by fusion, will also remove the organic portion. Most samples, however, must be oxidized first. Two general methods are available—reaction with oxidizing agents in solution (wet ashing) or with oxygen, either at elevated temperatures (dry ashing) or at room temperature with electrically-excited oxygen. When samples are heated, care is required to avoid loss of the desired constituents by volatilization and splatter. For example, samples to be analyzed for radioiodine cannot reliably be heated in the dry state but should be wet-ashed unless the iodine is in one of its more highly oxidized states. A selection of reported volatility losses is given in Table 5-1 and the references give additional data on other elements and conditions. The sample matrix and processing time affect the volatilization

TABLE 5-1—*Volatility losses on ignition*

Element	Matrix	Temperature	Time	Loss	Reference[a]
		°C	h	percent	
Polonium	Muscle	100	24	0	1
Polonium	Muscle	200	24	30	1
Polonium	Muscle	300	24	93	1
Cesium	Muscle	250	24	<2	1
Cesium	Muscle	400	24	<4	1
Cesium	Muscle	800	24	81	1
Polonium	Bone	150–250	24	5–10	1
Polonium	Bone	400	24	25	1
Lead	Bone	600	24	<1	1
Lead	Bone	700	24	21	1
Fission Product Mixtures (300–1300 days old)	Precipitation residues	500	—	2–3	2
		800	—	15–30	2
Manganese	Blood	400	24	1	3
Manganese	Blood	700	6	15	3
Cobalt	Blood	400	24	2	3
Cobalt	Blood	700	6	33	3

[a] References: 1—Martin and Blanchard (1969); 2—Gedeonov *et al.* (1972); 3—Korenman (1966).

rate of a given element. The chemical properties of the radioactive elements indicate when these problems will arise. Thus, metallic chlorides are generally more volatile than the corresponding sulfates, nitrates or carbonates. Chlorides of the alkali metals, lead, thallium, and other elements are nonvolatile. Other metallic elements of interest in environmental radioactivity measurements that can form volatile compounds under ordinary ashing conditions are technetium, ruthenium, mercury and zinc. Non-metallic elements are generally more volatile than metals.

Unexpected losses and discrepancies between results by various investigators have been reported. For example, Korenman (1966) refers to losses of 48 percent of iron from blood after ignition at 700°C for 6 hours, while other workers report little loss (Gorsuch, 1970). Dry ashing cannot be used for polonium, as Table 5-1 indicates, but Gorsuch (1970) states that this method has been shown to be quite satisfactory. Some of the reported losses may simply have been the result of conversion of the element to an insoluble form that made subsequent detection in the residue difficult, or may have resulted from mechanical entrainment. Regardless of cause, the analyst should be aware of the possibilities of loss on dry ignition of samples, and use the lowest practical temperatures.

Another reason for dry-ashing at the lowest feasible temperature is that some compounds become very difficult to dissolve after high temperature ignition, above 600–700°C. Since complete destruction of the organic matter is not always required, temperatures in excess of 500°C are not needed. An incompletely oxidized sample (gray ash) is acceptable for gamma-ray counting and many subsequent dissolution procedures. When igniting samples containing much organic matter, such as food, the temperature should be raised gradually over an extended period to avoid flame-burning and to minimize glow-burning. Losses of small particles by spattering can occur under these conditions. A partial oxidation of biological samples with nitric acid, followed by ashing in a muffle furnace allows a lower final temperature (400–450°C) and will oxidize some of the volatile elements to less volatile forms.

Since the ash obtained at low-oxidation temperatures is easier to dissolve than strongly ignited samples, wet oxidation is preferable if the sample is small, and is necessary if volatile activities are to be determined. This technique avoids most problems of volatility losses and insoluble residue formation, but is more difficult, requires more time and attention, and uses reagents for which a blank must be determined. Volatility problems still exist for some elements. Under acidic oxidizing conditions, iodine can be lost as elemental iodine, and some elements, such as ruthenium and technetium, are volatile from acidic solution in their highest oxidation state. Alkaline oxidizing agents (e.g., sodium peroxide) are required for samples intended for analysis of these elements. Otherwise acidic oxidizing agents, principally nitric acid, together with sulfuric and perchloric acids, are used. Concentrated hydrogen peroxide, 50 and 70 percent, is commercially available, and is very effective in hot nitric or sulfuric acid solutions. If sulfate can be tolerated in the chemical operations, the higher temperature obtainable with sulfuric acid gives better results. The sulfuric acid-catalyst (mercury) combination that is used in the Kjeldahl nitrogen determination has also been used for oxidation of organic materials for radiochemical analysis (Major *et al.*, 1964). If perchloric acid is used to complete the oxidation, it is essential that the destruction of organic matter be first carried nearly to completion with nitric acid, and hydrogen peroxide if needed, so that little organic matter remains before perchloric acid is added. Hot perchloric acid can react explosively with organic matter and with a few inorganic reducing agents (e.g., hypophosphites). It should be used carefully and in small quantities.

Wet ashing of large samples (0.1 to 1.5 kg) of materials such as meat, flour, fish, and dried milk with hydrogen peroxide-ferrous ion mixtures has been reported by Sansoni and Kracke (1971). The method appears promising and should find considerable use. The ferrous ion reacts with the peroxide to form hydroxyl radicals, which are presumed to be the effective oxidizing agent. The reaction is rapid; 100 g of meat can be reduced to a dry, white ash in 4 hours. Fats are not destroyed by this treatment and, if their removal is necessary, they can be separated physically or by solvent extraction.

Another method for decomposing organic material makes use of electrically-excited oxygen as the oxidizing agent (Gleit and Holland, 1962). The excited oxygen is passed over the sample at low pressures and oxidation occurs at 150°C or less. From the previous discussion, it is obvious that the method combines the best features of both wet and dry ashing. Equipment for this method is commercially available. However, the reaction rate is slow, oxidation occurs only near the surface so the sample must be stirred occasionally, only relatively small samples can be used, and the equipment cost per sample is high.

The discussion on ashing of environmental samples given by Harley (1972) is informative and should be consulted, as is the book on the destruction of organic matter by Gorsuch (1970). The latter work describes ashing methods useful for the determination of a large number of individual elements and gives a general discussion of most of the available ashing methods.

5.3.2.2 *Dissolution of Inorganic Residue*. Destruction of the organic material does not, of course, complete the sample preparation process if chemical separations are to be performed, and the inorganic ash must also be dissolved. Complete dissolution of the sample is not necessary if it can be shown that the desired nuclide is completely leached.

To dissolve residues, a variety of treatments with acids and fluxes is available. Since the methods used for environmental samples are similar to those used in conventional analytical chemistry, standard works on this subject, such as Hillebrand *et al.* (1953), should be consulted. The dissolution process will usually oxidize small amounts of organic material remaining after ashing, such as the "gray ash" referred to earlier.

Basic samples can be entirely, or almost entirely, dissolved in nitric or hydrochloric acids, or in their mixtures. Samples containing silica in reasonable amounts can be dissolved with hydrofluoric acid with the aid of nitric or hydrochloric acids. The fluoride can be removed by fuming with sulfuric or perchloric acids, or by complexing with aluminum ion or boric acid if they do not interfere with the subsequent analysis. An acid-decomposition, pressure method designed for the dissolution of samples for atomic absorption analysis may find application for radionuclide analysis, if the sample size can be increased (Bernas, 1968; 1973). At the present stage of development, 1-g samples of silicious inorganic material can be decomposed by heating for 30–60 minutes with hydrofluoric and nitric acids in a teflon-lined sealed container. Temperatures up to 170°C are used.

Substances not easily dissolved with acids must be fused with appropriate fluxes to obtain a solution. The higher temperature increases the reaction rates, but may vaporize some radionuclides. Insoluble metallic oxides or other metallic compounds can often be dissolved by fusion with the acidic compounds potassium or sodium pyrosulfate (or bisulfate, which on heating loses water to form the pyrosulfate). The resulting sulfate mixture can sometimes be dissolved in water, but dilute mineral acid is preferable, and the acid solution should be boiled for a few minutes to hydrolyze any condensed phosphates (which are good metal complexing agents) that have formed. Acidic compounds and many insoluble metal compounds, including silicates, can be attacked by fusion with sodium carbonate, sodium hydroxide, or a mixture of the two. The solidified melt can be dissolved in water (or dilute acid) and the silica removed by filtration if no insoluble carbonates are present. Otherwise the carbonates formed during fusion can be dissolved in mineral acid. If insoluble compounds can reform after dissolution of the carbonates in acid, the melt should be dissolved in water, the insoluble carbonates thoroughly washed to remove anions released during fusion, and the carbonates dissolved in mineral acid. In any procedure such as this, when more than one fraction is produced, all unwanted fractions must be checked for loss of the desired nuclide before discarding. Oxidizing fluxes (sodium nitrate, potassium nitrate, and sodium peroxide) can be used for samples to be analyzed for radioiodine, or the entire sample can be oxidized in a closed system (Studier *et al.*, 1962).

A combination of acids and fusion is often convenient because less material must be fused. As much as possible of the sample is first dissolved by mineral acids, including hydrofluoric acid, and the insoluble residue dissolved by fusion. A very rapid method for completely dissolving silica-containing materials such as soil consists of a preliminary nitric-hydrofluoric acid treatment, fusion with potassium fluoride, heating with sulfuric acid to drive off fluoride compounds and the addition of sodium sulfate and heating to produce a pyrosulfate fusion (Sill, 1961). The melt is completely soluble in dilute hydrochloric acid, except for about a milligram of barium sulfate. The procedure can be completed in 30 minutes, but has two limitations. It is difficult to dissolve completely more than 50 g, although in some situations samples as large as 0.1 to 1 kg may be necessary to obtain a representative sample and sufficient activity for measurement. Also, calcium sulfate may precipitate from samples high in calcium and subsequent chemical separations must be compatible with this condition. The sulfate medium has the advantage that some elements of interest, such as plutonium, form complexes with sulfate ion that help keep them in solution.

The container in which samples are dry-ashed or fused requires consideration. For fusions, platinum is the most versatile, but the most expensive, and cannot be used with hydroxides or alkaline oxidizing agents. Such fusions may be carried out in iron or nickel crucibles, which are attacked but replaceable at reasonable cost. Ashing is best done in platinum, but borosilicate glass is a suitable substitute at temperatures below 500°C provided it is not attacked or etched. Vycor, quartz, or porcelain can be used at high temperatures in place of platinum, but they are subject to breakage, are not inexpensive, and are often etched by the sample, after which cleaning is difficult. For samples larger than about 500 ml, platinum becomes prohibitively expensive, but stainless steel and monel containers are suitable up to 500°C. The sample may become contaminated with traces of the metals, but this usually causes no problem. Some laboratories have found that ceramic containers designed for household use are satisfactory.

The dissolution procedure should also include some step to insure that any added carrier or tracer is in the same chemical form as the nuclide being analyzed. That is, isotopic exchange between the two should be insured by a treatment that depolymerizes colloidal plutonium or dissolves refractory plutonium oxide (fluoride ion or pyrosulfate fusion), complexes protactinium or zirconium (fluoride ion), converts multivalent elements to their highest or lowest oxidation states (hypochlorite oxidation of iodine or ruthenium, sulfite reduction of plutonium), or complexes or dissolves radium and barium sulfates (diethylene triamine pentaacetic acid (DTPA) or sodium carbonate fusion). For many nuclides, dissolution in mineral acids is sufficient to effect exchange. Attention must also be given to the chemical and physical form of the radionuclide aside from the isotopic exchange question, even when a carrier or tracer is not used. The radionuclide must be brought into a form in which it undergoes the expected reactions in the separation process.

5.3.3 *Air Filter Samples*

Samples of particulate matter in air intended for total-activity counting require little preparation other than that needed to obtain a reproducible and suitable size, such as folding or cutting if the sample area is much larger than the detector. Most laboratories that have reported their procedures for this type of measurement do not specify any sample preparation, except to place them in plastic envelopes for shipment and storage. If there is concern for the loss of material from dust-covered paper, the paper can be folded on itself for shipment and for gamma-ray counting. To fix the dust particles on the paper so the surface can be exposed for alpha and beta counting and so the samples can be easily handled without loss, the surface can be covered with a thin film. For this purpose, a solution of a plastic, such as polymethylmethacrylate, dissolved in a volatile organic liquid such as ethylene dichloride, can be sprayed on the sample with a low-velocity atomizer to a thickness of about 0.1 mg cm^{-2}.

If uniformity of activity is to be insured in thick, dust-loaded filter papers, or if the paper permits penetration of the particles, a treatment such as ashing and grinding the ash is needed. This is important for alpha emitters since the range of alpha particles in a light element matrix such as soil is only about 5 mg cm^{-2}, or 0.02 mm if the density of the material is 2.5 g cm^{-3}. At low beta energies the same considerations apply as for alpha particles. A 0.05-MeV beta particle has about the same range in light elements as a 5-MeV alpha particle. To minimize the penetration problem, samples can be collected on membrane filters. These offer, in addition to collection of the particulate matter almost entirely on the surface with little penetration, nearly complete collection efficiency. They have the disadvantages of high cost and high pressure drops.

Most of the organic matter in an air filter sample is in the filter medium, and the oxidation methods depend on its nature and on the radionuclides sought. The general information in Section 5.3.2.1 applies. Ignition is the easiest method, but oxidation of membrane and polystyrene filters with nitric acid and hydrogen peroxide is not difficult.

After removal of the organic material, the sample itself must be dissolved. The sample is essentially a small soil sample, and can successfully be treated as such. A selection of procedures suitable for dissolving air filter samples is given in Table 5-2.

5.3.4 *Water*

The general problem of the measurement of radioactivity in environmental water has been discussed by Kahn (1972). The preparation of water samples has the purpose of preserving the integrity of the sample during storage. The two principal pretreatment methods are acidification and filtration. The addition of carriers or complexing agents, or freezing, may be effective for some situations. The purposes of the radioactivity measurements should determine what, if any, pretreatment is used. Within the broad categories of measurement for dose assessment or legal compliance purposes, or for research, it may be desirable to measure total activity in water, the distribution of activity between water and sediment, the volatile nuclides, including tritiated water, or the chemical form of a radionuclide (colloidal, ionic, complexed, etc.). In most situations the type of preparation is obvious once the purpose of the measurement is established. However, it may be difficult to define the details of the pretreatment, and factors to be considered are discussed below.

The preparation of environmental water samples for analysis presents a number of problems not shared by other sample types and for which clearly defined answers are not available. The problems are caused principally by two situations; the samples are two-phase systems (i.e., the water contains some insoluble material), and the behavior of many substances at very low concentrations is unpredict-

TABLE 5-2—*Procedures for dissolving air filters for radiochemical analysis*

Filter Medium	Radionuclides	Method	Reference[b]
Polystyrene fiber; membrane	Plutonium and other metallic radionuclides	Heat with nitric and hydrochloric acids, evaporate; heat with hydrogen peroxide; heat with nitric and hydrofluoric acids, fume with perchloric acid.	1
Not specified	Plutonium, Strontium, Cesium	Ignite at 425°C in muffle furnace; treat with nitric, hydrofluoric, and perchloric acids; fuse any insoluble portion with sodium carbonate.	2
Glass fiber	Plutonium, Uranium, Strontium	Fuse with sodium carbonate, dissolve in hydrochloric acid, treat insoluble residue with hydrofluoric and hydrochloric acids.	3
Cellulose	Plutonium, Uranium, Strontium	Ignite in muffle furnace,[a] dissolve residue in hydrofluoric and hydrochloric acids.	3
Membrane filter paper	Plutonium, Uranium, Strontium	Dissolve paper in acetone, evaporate to dryness, ignite[a] residue, dissolve in hydrofluoric and nitric acids.	3
Polystyrene, membrane, cellulose, or glass fiber	Heavy element alpha emitters	Ignite with burner or oxidize with nitric, sulfuric, and perchloric acids; fuse residue with sodium fluoride, remove fluoride by heating with sulfuric acid, add sodium sulfate and fuse. Dissolve in hydrochloric acid.	4

[a] Temperature not specified, but 600°C should be sufficient.

[b] References: 1—Harley (1972); 2—Guthrie and Grummitt (1963); 3—Douglas (1967); 4—Sill and Williams (1971).

able or poorly understood. The variable composition of natural waters adds to the problem. The question of how, if at all, to treat samples between collection and analysis has not been unequivocally resolved, and there is little experimental evidence on which to base a decision. Additional work remains to be performed on this problem, and the preferred treatment may vary with the particular body of water sampled and with the specific analyses to be done.

It is known that many cations at the very low concentrations of interest in the environment can be lost from solution or no longer exhibit their expected ionic behavior. This change is most likely to occur at low acidities, and may be due to hydrolysis and sorption on the container walls or on particles in the water. Even an ion as simple as cesium can be lost in trace amounts in glass containers by exchange with the potassium in the glass. Radionuclides at low acidities may exhibit many of the properties of colloids, and the term "radiocolloid" has been applied to such mixtures. Whether a radionuclide at a concentration of about 10^{-15} M can form a true colloid (when its solubility product has not been exceeded) or whether the nuclide is part of, or adsorbed on, another colloidal particle is immaterial. One should, however, have an appreciation of the very low concentrations of interest and of the unusual behavior that can occur. For a discussion of radiocolloids in general, see Bonner and Kahn (1951) and Crouthamel and Heinrich (1971). The formation and characterization of plutonium colloids have been studied by Lindenbaum and Westfall (1965) and Andelman and Rozzell (1970).

While it is important to avoid losses to container walls during storage, it is also important that the pretreatment not induce a change in the distribution of the radionuclides between the water and the particulate matter, or induce volatility losses of radionuclides. Thus, to reduce the possibility of the loss of radionuclides on the walls, the sample is usually acidified. Most environmental water samples are close to neutral and acidification to 0.1 to 0.5N is adequate. At this acidity, radioactive substances in particles in the sample can be leached (perhaps slowly) and appear in the solution. The resulting increase in the original soluble radioactivity can be considerable. Also, if nitric acid is used, it will slowly oxidize iodide ion to iodine, and radioiodine in solution may be lost by volatilization. Consequently, it would appear that if acidification is desirable to prevent hydrolytic losses of radionuclides, the sample solution should first be filtered. The acidity during filtration and the choice of filter medium are also important, since considerable loss may occur during this process. An efficient method for separating traces of actinide elements from solution utilizes filtration through glass-fiber paper at pH 5 (Eakins and Gomm, 1968). Most actinides are absorbed almost completely on the glass fibers, an indication that silica in solution can adsorb many ions from solution. Reynolds (1962) reports that filtration through ordinary filter paper of a solution containing a fission product mixture resulted in the recovery of a very small precipitate that contained appreciable quantities of all the fission products except cesium. The distribution of radionuclides in seawater between solution and particulate matter also remains an open question, since the reports by different investigators are contradictory. Silker (1972) has been investigating this problem, and has found that generally less than 10 percent of ^{7}Be and several fission products was present in particulate form, although occasional samples ranged up to about 40 percent of the activity as particles.

On the basis of the foregoing discussion it would appear that the safest pretreatment procedure, if the filterable activity and particulate activity are desired separately, is to filter the water through membrane filter paper while collecting the sample in the field, and acidify the sample immediately thereafter. The acid can be present in the receiving flask during filtration. In this way the water is not in contact with the container walls until after acidification (Olson, 1973). If this procedure is impractical, the sample should be filtered as soon as possible after collection; then it is safer to acidify the sample immediately prior to filtration. Centrifugation can be used to avoid the uncertainties of filtration if the samples are not prohibitively large.

The sample can also be treated in such a way as to combine insoluble and soluble forms of the nuclide to obtain total activity. For nonvolatile substances the sample can be acidified, evaporated to dryness, and the residue dissolved and analyzed (Harley, 1972; Guthrie and Grummitt, 1963). Any inactive carriers or radioactive tracers used in the analyses should be added as soon as possible, preferably before acidification, filtration, or evaporation. The U. S. Public Health Service (Douglas, 1967) usually calls for filtering the sample (unless it is potable water, in which case the total activity is desired) without acidification to obtain the distribution of activity. In this case, the asbestos filter suggested as one possible filtering medium does not appear as safe as the membrane and cellulose papers also suggested.

An important consideration in radiochemical analysis of potable water for radiological health purposes is a determination of what information is needed for calculations of the dose to man from drinking water. Very insoluble activities, such as small particles of insoluble plutonium oxide, will not dissolve readily in the GI tract and thus will contribute little in the way of systemic dose, and in this respect should not be included as radioactivity in water, although they would irradiate the intestine. One solution to this problem is to acidify the sample to simulate the acidity conditions in the stomach, and then filter (Porter *et al.*, 1965). The analysis of both fractions should lead to results of dosimetric significance.

There are some samples, however, for which acidification will not prevent sorption of activity on container walls. Olson (1973) reports considerable loss of radioactive nuclides on walls of polyethylene bottles containing reactor effluent water. This loss was not preventable by prior acidification, nor could the sorbed activity be readily removed, except by wiping with filter paper. It is believed that the nuclides are absorbed by algae growing on the walls. To obtain a sample for analysis that contained this activity, a thin polyethylene bag was placed in the bottle before adding the sample, and the bag was analyzed with the water.

If a sample has been stored before analysis, particularly at its natural acidity, it is advisable to transfer the sample to a second container and wash the original container with a solution—usually an acid—in which the sought-for activity is soluble. The effectiveness of the washing procedure must be verified experimentally. The wash solution can be added to the original solution or analyzed separately. Freezing may be a suitable method for preventing sorption of radionuclides on container walls between sampling and analysis, but is practical for small samples only, and the treatment of the insoluble portion still remains.

A precaution to be observed in gamma-ray counting of water samples is the change in counting geometry that will occur if finely divided suspended matter containing activity settles out or if soluble radioactive species become fixed on the container walls during counting. Unless difficulties such as those observed by Olson (1973) occur, such samples can be acidified, if the suspended material is soluble in dilute acid, filtered, or evaporated prior to counting. An aliquot from a larger sample of this type of mixture must be removed carefully. Suspended material should be included in the sample by mixing the solution vigorously during sampling.

While standard rules are not given for treating water samples, this discussion of the difficulties and problems that can occur should alert the analyst and allow him to preserve, treat, and prepare the sample in a way that will give the desired results. Additional studies in this area are undoubtedly needed.

5.3.5 *Solid Samples*

The preparation of soil, sediments, vegetation, food, biota, and similar materials for analysis does not present the same problems described for water samples, and in this respect is simpler although more laboratory operations, equipment, and time are needed. One important aspect of the analytical process is the moisture-content basis on which the results are calculated. The basis should correspond to the way the results are to be used, e.g., for dose assessment, human diet constituent results should be reported on a wet basis. Soil and vegetation can be dried at room temperatures. Oven drying is probably more reproducible, but may be impractical for very large and bulky samples. One solution is to dry the sample at room temperature and after mixing, dry a small portion in the oven to determine the moisture content.

Oven drying temperatures can range from 80–140°C. If the results are to be calculated on the basis of oven-dried sample weight, samples sensitive to thermal decomposition, such as food and vegetation, should be dried at the lower end of this temperature range. Freeze drying is an excellent, although expensive, method for drying large samples containing considerable organic matter in preparation for ashing. This technique avoids the spattering and frothing that occurs during heating.

After samples such as soil and vegetation have been dried, they should be reduced to a small particle size by grinding and made homogeneous by mixing. Commercially available grinders, mills, pulverizers, mixers, and blenders are suitable. Hardened metals are best for the grinding surfaces, and it is not likely that radioactive contamination will be introduced into a sample with this equipment, except by cross-contamination from a high radioactive sample. To prevent this, thorough cleaning of equipment between samples is advisable. Material such as food and certain biological specimens can be reduced to a homogeneous condition for proper sampling with an electric food-type blender. At this stage a portion may be taken for gamma-ray spectrometry, either of the ground and mixed soil or vegetation sample or the semiliquid slurry of food samples. Large tissue samples are best oxidized to remove organic material, then mixed and sampled for gamma-ray spectrometry and radiochemical analysis.

A detailed procedure for preparing soil samples for analysis is given by Harley (1972). Before crushing and blending, large rocks should be removed, weighed separately and discarded. The sample is then processed. When synthetic nuclides are to be determined, it can be assumed that the large rocks contain an insignificant fraction of the activity and are, in effect, voids in the sample. This has been shown experimentally for plutonium (Krey and Hardy, 1970). The radionuclide concentration in the soil fraction can be converted to the concentration in the original sample by correcting for the volume occupied by the rocks. The method used in expressing the results should be stated. The results, in terms of deposition (concentration per unit area), will be the same in either case provided the area sampled is measured.

After the sample has been dried, weighed and mixed, it is ready for the oxidation and dissolution treatments described in Section 5.3.2. Among the available treatments, the choice is frequently one of personal preference. There is general agreement that vegetation, food, and tissue samples require complete dissolution. The question of complete dissolution versus leaching of soils is still being debated for certain analyses. It is believed that ^{90}Sr can be determined with equal accuracy by complete dissolution or by proper leaching procedures (Harley, 1972). Plutonium from world-wide fallout can be successfully leached with acids if a plutonium tracer in the same chemical form is used to monitor the yield. Other forms of plutonium (such as occur in fused particles near the location of a nuclear test or are produced in a laboratory or fuel reprocessing plant by high temperature ignition) require the fluoride-pyrosulfate or other fusion methods or repeated treatments with hydrofluoric and mineral acids for complete dissolution. However, little experience with these forms of plutonium in the environment is available. The proceedings of a symposium on environmental plutonium contain some information on this question (Fowler *et al.*, 1971). A selection of acceptable procedures for solid samples is given in Table 5-3.

5.3.6 *Milk*

The evaporation of milk to dryness prior to ashing causes difficulty because of the formation of a protein skin on the surface. Many analysts avoid this problem by separating radionuclides from the original liquid sample by techniques such as ion exchange. If evaporation is used, problems can be minimized by operating at a low temperature with stirring, by adding milk in small quantities to the container in which evaporation occurs so only a small quantity is being heated at any one time, by coagulating the protein before evaporation by heating the milk with rennin, or by freeze-drying. Ashing and dissolution of the inorganic residue is accomplished by the methods described in Section 5.3.2. Table 5-4 lists suitable methods. The preservation of milk samples is discussed in Section 5.2.4.3.

5.4 Radiochemical Separations

5.4.1 *General Considerations*

5.4.1.1 *Precautions in Separations at Very Low Concentrations.* The principles and kinds of radiochemical separations useful for

TABLE 5-3—*Procedures for ashing and dissolving environmental samples*

Sample type	Radionuclides	Method	Reference[a]
Soil (0.1 to 1 kg)	Plutonium (fallout)	Leach with HCl-HNO_3—^{236}Pu tracer for yield monitor.	1
Soil (5 g)	Plutonium	HF-HCl—fuse with KF, $K_2S_2O_7$—dissolve in HCl.	2
Soil (10–50 g)	Plutonium, other actinides	HF-HNO_3—fuse with KF, heat with H_2SO_4 and Na_2SO_4—fuse—dissolve in HCl.	3, 4
Soil (100 g)	Plutonium	Ash 450°C—leach with 6M HCl by refluxing overnight.	5
Foods (0.1–4 kg)	Strontium-90	Dry (150–225°C) and ash, first at 150–325°C, then at 500°C, fuse with Na_2CO_3—dissolve in water, filter, dissolve carbonates in HNO_3.	6
Vegetation (500 g)	Various non-volatile radionuclides	Dry 105°C, ash 550°C—dissolve in concrete HCl, discard insoluble residue.	7
Soil (100 g)	Strontium-90	Ignite 400–500°C—fuse with Na_2CO_3 at 900°C, dissolve with HCl, filter silica (complete dissolution method).	6
Soil (250 g)	Strontium-90	Add strontium carrier, ^{85}Sr tracer, heat with NaOH, then heat with HCl, then HNO_3 (leach method).	6
Vegetation (20 g)	Radiodine	Add carrier, oxidize with ceric, permanganate, H_2SO_4; reduce iodate to I_2 with oxalic acid-sodium hydroxide and distill iodine.	8, 9
Vegetation (100 g)	Radioiodine	Add carrier, fuse with NaOH and KNO_3, dissolve in water, filter.	10
Thyroid	Radioiodine	Add carrier, fuse with NaOH and KNO_3, dissolve in water, filter.	8, 9

[a] References: 1—Chu (1971); 2—Butler *et al.* (1971); 3—Sill and Williams (1969); 4—Sill *et al.* (1974); 5—deBortoli (1967); 6—Harley (1972); 7—Douglas (1967); 8—Technical Manager (Chemistry) (1961); 9—WHO (1966) 10—Morgan and Mitchell (1962).

environmental samples are the same as those for any type of material. However, additional attention must be paid to a number of important details. At the very low concentrations frequently encountered, contamination of the sample with extraneous radionuclides during processing must be avoided, and the blank value must be known and kept low and constant. Reagents must be carefully checked for radioactivity. Glassware, other laboratory equipment,

TABLE 5-4—*Procedures for preparing milk samples for analysis*

Radionuclides	Method	Reference[a]
Various nonvolatile radionuclides	Add rennin—evaporate to dryness, dry at 140°C—ash progressively at 250°C, 425°C, and 525°C, dissolve in conc. HCl, discard residue.	1
Various nonvolatile radionuclides	Milk is pumped at 5–8 ml min^{-1} into rotating evaporating dish heated by blast burner; the system evaporates 1 liter to a charred mass in 2.5–3 hr; ashing is completed at 450°C for cesium and at 550–600°C for other nuclides.	2
Radioiodine	Add iodide carrier; NaOH, evaporate to dryness, ignite at 500°C, fuse with KNO_3 and NaOH, dissolve cooled melt in water.	3, 4
Radiostrontium	Evaporate to dryness at 125°C; heat slowly (8 hours) from 175 to 325°C; ignite at 485°C, add carrier, dissolve in HNO_3, filter and discard residue.	5

[a] References: 1—Guthrie and Grummitt (1963); 2—Murthy and Campbell (1959); 3—Technical Manager (Chemistry) (1961); 4—WHO (1966); 5—Harley (1972).

and the laboratory itself, must be kept radioactively clean. In low-level radiochemical analysis, the analyst should recognize that frequent blank and background determinations are as important as the sample analysis. In general, blank and contamination problems are minimized by using the least number of reagents and separation steps possible. In selecting reagents, it is possible to make profitable choices. For example, an organic base (as ethanolamine) is less likely to contain radionuclides, such as ^{226}Ra, than inorganic bases (as sodium hydroxide); calcium, as a carrier for actinide elements, is less likely to contain thorium than is lanthanum; and for neutralizing water prior to tritium enrichment by electrolysis, sodium peroxide or carbonate is preferable to sodium hydroxide. In addition to a low blank, a high radiochemical purity in the separation is also needed for good accuracy and sensitivity. For example, the environmental concentration of thorium is approximately 10^3 times that of the chemically similar plutonium, so a decontamination from thorium of a factor of 10^5 or more is required for a satisfactory plutonium analysis.

The extremely low concentrations of radioactive nuclides in the environment must be considered in devising workable radiochemical separation procedures. An ^{131}I concentration in milk of 1 pCi l^{-1} is approximately 6×10^{-20}M and represents only about 3.7×10^4 atoms. The concentration of ^{239}Pu in surface or ocean water from fallout is of the order of 5×10^{-4} pCi l^{-1}, or about 5×10^{-17}M, and this concentration is obtained with 2×10^{7} atoms. In soil, a common plutonium concentration from fallout is about 0.1 pCi g^{-1}. This activity can be obtained from one plutonium oxide particle with a diameter of 0.1 μm. The behavior of substances at these levels may not be predictable from their behavior at ordinary concentrations since the chemical species may be different (Bonner and Kahn, 1951). At very low concentrations, adsorption on impurities and particles and by microorganisms can be significant, or even complete, and the "radiocolloid" problem discussed in Section 5.3.4 in relation to natural water samples must be considered in all solutions in the course of an analysis, and particularly in the long-term storage of radioactive standard solutions. The latter are frequently sterilized or preserved with formalin to prevent microorganism growth.

These difficulties due to very low concentrations can be avoided by adding macro quantities of inactive isotopes of the same element, or of similar elements if no stable isotope exists. An example of the latter is the use of barium in radium separations. This procedure, the use of "carriers" for radioactive tracers, is frequently employed for measuring the chemical yield of a separation procedure. The chemical separations are greatly simplified since now the element can be expected to exhibit its normal behavior, and the analytical problem is no longer one in ultra-trace chemistry. For this reason it may be advisable to add an inactive carrier even when it is not used to measure the chemical yield. Attention must also be given to assuring that the radionuclide is in the desired chemical and physical form, that is, in proper ionic or non-ionic form and in the proper oxidation state to undergo the expected chemical reactions (see discussion on plutonium in Section 5.3.5).

An example of the problems that have been encountered is the following. Sill and Williams (1969) and Sill and Willis (1964) have found that barium and other insoluble sulfates co-precipitate the tri- and tetravalent actinides quantitatively under the proper conditions. The few milligrams of barium and strontium naturally present in 10 g of soil can produce serious losses of these elements when soil is dissolved in a sulfate system and the small quantity of insoluble sulfate discarded.

5.4.1.2 *Measurement of Chemical Recovery*. The chemical recovery in a radiochemical separation must obviously be known to obtain quantitative results. This problem is common to all trace analyses. While ordinary quantitative analyses are designed to result in complete recoveries, this is not always feasible in radiochemical analysis, particularly of complex samples such as environmental materials, nor is it necessary if the recovery can be measured for each sample. Two general methods are used for this purpose. One is to measure the yield in similar or identical samples by adding a known amount of the nuclide sought, and applying this recovery figure to the unknown sample. This method has obvious disadvantages and uncertainties, and should be used only if better methods are not available. If this method is used, the most favorable condition is to analyze duplicate portions of the same sample simultaneously and add a known amount of the nuclide sought to one of the aliquots to obtain the recovery for that sample.

The second method involves adding a known quantity of an isotope of the element sought which can be measured after the chemical separation is completed. This gives the yield directly and is preferred if both isotopes can be measured without significant, or with correctable, interference from each other. The technique has much in common with the isotope dilution method used in mass spectrometric analyses for many years. The added isotope may be radioactive or inactive. If an inactive isotope is available, it is common to add a weighable quantity to the sample, usually 20 mg, and separate the element in a weighable form to obtain the yield. The separated material can usually be counted after weighing without additional operations. For accurate results requirements are as follows: a reproducible weighing form, a pure (but not necessarily quantitative) separation, and an independent measurement of the carrier element if it is a natural constituent of the sample. The latter consideration is significant when, for example, large samples of soil (100 g) are analyzed for radiostrontium and large milk samples (4 l) for radioiodine. The fission product analytical procedures developed by the Manhattan Project (Coryell and Sugarman, 1951) all use inactive carriers and measure the recovery by weighing. Gravimetric yield determinations are not necessary, and any quantitative measurement (colorimetry, electroanalysis, atomic absorption spectrometry) can be used.

Radioactive isotopes can also be used to measure recovery where they can be distinguished from the nuclide sought through their decay properties. For elements with no stable isotopes, this method is, of course, required, and even when stable isotopes are available, a radioactive tracer can frequently be used to simplify the analysis. For example, ^{85}Sr tracer is often used to obtain the strontium chemical yield by gamma-ray counting in analyses for ^{90}Sr, even though inactive strontium carrier is added to perform the radiochemical separation by precipitation reactions. This tracer can be used when ^{90}Y is separated and counted to obtain the ^{90}Sr parent activity. It cannot be used when the ^{90}Sr itself is counted, or when ^{89}Sr is to be determined, because although ^{85}Sr decays by electron capture, it gives sufficient response to interfere with the beta counting of ^{90}Sr or ^{89}Sr. For ^{90}Sr determinations, ^{85}Sr has these three advantages:

(a) The yield can be measured simply by gamma-ray counting of ^{85}Sr in a solution of the separated strontium fraction without precipitating and weighing the strontium carrier. The strontium is then available for the separation of ^{90}Y without the additional steps of redissolving the strontium precipitate.

(b) The natural strontium content of environmental materials need not be determined, because it does not affect the accuracy of the yield determination.

(c) The separated strontium fraction need not be free from calcium.

An examination of the Table of Isotopes (Lederer *et al.*, 1967) or other compilations of nuclear decay data will indicate which nuclides can be used as isotopic tracers for chemical recovery. Availability sometimes limits the choice to less desirable nuclides. Some are available commercially or from laboratories of the U. S. Energy Research and Development Administration, while others must be made by the analyst or routinely separated from a longer-lived parent. A selection of suitable nuclides is given in Table 5-5.

TABLE 5-5—*Radionuclide tracers suitable for measuring chemical recovery*

Radionuclide	Tracer Nuclide	Method of Measurement	Remarks
^{90}Sr	^{85}Sr	γ counting	Usable when ^{90}Y is counted.
^{99}Tc	^{99m}Tc	β or γ counting	^{99}Tc counted after ^{99m}Tc decays; ^{99m}Tc obtained by separation from ^{99}Mo parent.
^{147}Pm	^{149}Pm	β or γ counting	^{147}Pm counted after ^{149}Pm decays, produced by ^{148}Nd (n, γ).
^{226}Ra	^{225}Ra	β counting	Separated from ^{233}U decay series.
^{230,232}Th	^{234}Th	β counting	Separated from ^{238}U parent.
^{237}Pa	^{233}Pa	β, γ counting	Produced by ^{232}Th (n, γ).
^{235,238}U	^{233}U	α spectrometry	Produced by ^{232}Th (n, γ)
^{237}Np	^{239}Np	β counting	Separated from ^{243}Am.
^{210}Po	^{209}Po	α spectrometry	Produced by ^{209}Bi (p, n)
^{238,239}Pu	^{236,242,244}Pu	α spectrometry	^{242,244}Pu also measurable by mass spectrometry
^{241}Am	^{243}Am	α spectrometry	
^{252}Cf	^{249}Cf	α spectrometry	

The use of alpha emitters to monitor the chemical yield of other alpha emitters places severe requirements on the chemical separation procedure and on the isotopic and radiochemical purity of the tracer. The effect of the isotopic purity of the tracer on the sensitivity of the analysis, and on the amount of tracer that can be used, which determines in part the accuracy of the analysis, is apparent. Similar considerations apply to the effect of radiochemical purity of the tracer on sensitivity and accuracy when, for example, several actinides are analyzed in tandem in the same sample, and tracers are used for each actinide. The desired nuclide must be separated free from inert solids as well as interfering activities. This is necessary whether the final measurement is made by alpha particle spectrometry or mass spectrometry. Some of the considerations that affect the choice of the tracer isotope are brought out in comparing the relative merits of ^{236}Pu, ^{242}Pu and ^{244}Pu as monitors for plutonium to be measured by alpha spectrometry. The former has a relatively short half-life (2.85 y) and has a large number of alpha-emitting daughters, making absolute standardization difficult. Its alpha particles have energies higher than those of ^{238}Pu and ^{239}Pu, and, since weightless sources are impossible to prepare, this limits the amount of ^{236}Pu that can be added to the sample without interfering with the other plutonium peaks by energy degradation in the source. However, ^{236}Pu free from other plutonium isotopes can be prepared from pure ^{235}U. In most cases, better tracers for plutonium are ^{242}Pu and ^{244}Pu. Both have a very long half-life ($>10^5$ y), daughters accumulate very slowly, and since their alpha particle energies are less than those of ^{238}Pu and ^{239}Pu, larger quantities can be used without causing interference. Their half-lives are so long, however, that they are difficult to prepare free from other plutonium isotopes on an activity basis. Electromagnetic isotope separation is necessary to obtain pure ^{242}Pu or ^{244}Pu. When the ^{239}Pu content of a group of samples is unknown or varies widely, the ^{239}Pu can obscure the small amount of ^{242}Pu or ^{244}Pu tracer in an alpha spectrum. In such cases, ^{236}Pu is preferred.

When isotope enrichment is used to increase the sensitivity of measurement, the recovery can often be measured with naturally-occurring isotopes. For ^{3}H and ^{14}C, the enrichment can be determined by measuring ^{2}H (by infrared absorption) and ^{13}C (by mass spectrometry) since these isotopes will follow the ^{3}H and ^{14}C closely in the isotope separation process.

Non-isotopic tracers can also be useful to monitor chemical yields with adequate reliability if the chemical properties are very similar. Thus, if the chemical losses are low, ^{133}Ba can be used for radium, a short-lived neodymium isotope for promethium, and americium and curium for each other.

Whatever method is used to measure the recovery of a chemical separation procedure, isotopic exchange, or chemical and physical equivalency between the radionuclide in the sample and the added carrier or tracer is essential. This has been discussed briefly in Section 5.3.2.

An isotopic exchange should be undertaken, as early as possible in the analysis, before any chemical or physical operations are performed that might result in loss of the desired nuclide. It is often necessary to make a compromise between the most scientifically satisfying time to add tracer and effect exchange, and the earliest feasible time. This occurs, for example, in the analyses of foods and vegetation, and for solids in general.

5.4.1.3 *Problems in Attaining Isotopic Exchange*. Some additional comments on exchange are appropriate here. In the case of the simpler uni- and divalent ions, placing the sample in a medium in which the element is soluble is sufficient. For cesium and strontium, an acid solution is adequate. However, if some of the ^{90}Sr is tightly bound in a solid, such as fallout strontium in soil, complete dissolution of the soil by fusion or the addition of strontium carrier followed by an alternate basic-acidic treatment was found necessary to obtain complete exchange (Harley, 1972). Otherwise only that fraction of the ^{90}Sr solubilized by the treatment will be recovered in the same yield given by the carrier or tracer.

The situation is more complicated for the more highly charged ions and for those than can exist in many valence states. In the latter case, the sample should be treated in such a way that all forms of the element will be converted to the same oxidation state before exchange can be assumed. A good example of this is found in the analysis of fission product iodine (Coryell and Sugarman, 1951) where it was found that exchange between the iodide carrier and fission product form of the iodine was uncertain until all of the iodine was carried to its highest oxidation state. The first steps in the analytical procedure then, were to add iodide carrier, make the solution alkaline, and then heat it in the presence of hypochlorite to carry the iodine through all its valence states to the periodate. A similar situation was found for ruthenium in its separation from technetium (Golchert and Sedlet, 1969).

Elements that form strong complex ions can be converted to that form to obtain exchange and place all forms of the element in a state with reliable, predictable chemical behavior. The fluoride complexes of zirconium and protactinium can be used for this purpose.

The tetravalent and pentavalent metal ions, as plutonium, protactinium, zirconium, thorium, and polonium tend to hydrolyze in weakly acidic solution, and when present in trace concentrations, can form large non-ionic species (the "radiocolloids" or "polymers" discussed previously). In this state they will not exhibit their expected ionic behavior in chemical separations such as ion-exchange. Complex formation with fluoride ion has been mentioned as a means of converting some of these to ionic forms. Other methods to accomplish this are prolonged heating in strong mineral acids, fusion with pyrosulfate, and presumably, changing the oxidation state since the tri- and hexavalent oxidation states will exist as simple ions in less acidic solutions than will the (IV) and (V) states. Plutonium is an important environmental nuclide, and since it is most stable in the tetravalent state, some treatment is necessary to insure it is in ionic form. The following treatments are believed to perform this function: fusion with potassium pyrosulfate, reduction to the trivalent state and treatment with fluoride, and heating with strong mineral acids, preferably including hydrofluoric acid.

5.4.2 *Separation Methods*

5.4.2.1 *Survey of Techniques*. A large number of separation methods have been applied in analytical radiochemistry with considerable success, and all of these methods have been used for environmental analysis. Their success has been remarkable when one considers that a relatively small number of atoms must be separated in a pure state from a very large amount of material.

The principles and details of the separation procedures will not be presented in this report. For this information the literature should be consulted, and the references cited below will give the analyst a good background in the methods and additional references to most of the literature in the field. For good general discussions on radiochemical separations and analyses, the monographs by Lavrukhina *et al.* (1967); Wahl and Bonner (1951); Crouthamel and Heinrich (1971); McKay (1971); and the manual by Harley (1972) should be consulted. Excellent sources of information on specific separation methods are: coprecipitation (Wahl and Bonner, 1951; Crouthamel and Heinrich, 1971; Vdovenko, 1967); liquid-liquid or solvent extraction (De *et al.*, 1970; McKay *et al.*, 1965; Morrison and Frieser, 1957; Marcus and Kertes, 1969; Moore, 1960; Peppard, 1971); ion-exchange between solid and liquid phases (Marcus and Kertes, 1969; Massart, 1971, Kraus and Nelson, 1957; Lavrukhina *et al.*, 1967); chromatographic separations—paper, gas, gas-liquid, electrical, and other (Bailey, 1962; McKay, 1971; Crouthamel and Heinrich, 1971; Schumacher, 1957); volatility (Crouthamel and Heinrich, 1971; DeVoe, 1962); electrochemical (Wahl and Bonner, 1951; Crouthamel and Heinrich, 1971; McKay, 1971); foam separations (Lemlich, 1972; Crouthamel and Heinrich, 1971); and isotopic exchange reactions (Crouthamel and Heinrich, 1971; Wahl and Bonner, 1951). Precipitation (and co-precipitation), ion exchange, and liquid-liquid extraction methods have been the most widely used methods for environmental samples, the first because of its wide applicability and the last two because the separation processes usually function well at the very low radionuclide concentrations of interest, and do not depend on mass. Electrochemical separations have had their widest application in the uniform deposition of separated radionuclides on metals to prepare uniform counting sources. In this application, separation from impurities is not the principal function. Many applications of volatility separations readily come to mind for the noble gases, iodine, ruthenium, and other elements that form volatile compounds. Among the separation methods that have not been widely used for environmental samples, isotopic exchange, foam extraction, and partition chromatography should be particularly useful but have not yet been fully exploited.

In searching for a separation procedure, one should first consult the manuals issued by several environmental analysis laboratories and referred to in Section 5.1 to determine if a directly applicable method is available. In the study of new methods, a prime source of information on the radiochemical properties of the elements and on radiochemical techniques is the series of monographs written by various authors under the sponsorship of the National Academy of Sciences and National Research Council (NAS-NRC, 1960–1971). The radiochemistry of all but a few of the elements (hydrogen, lithium, boron, thallium, bismuth, and neptunium) through nobelium is covered in a series of 58 monographs, each dealing with a single element or group of related elements. Also, a number of radiochemical techniques (separations, counting, etc.) are covered in 13 additional monographs. Although the methods given in this series are not directed toward environmental samples, they can usually be adapted for such use. One caution in their use is the age of some of the monographs. The earliest ones were issued in 1960. Some have not been revised for many years, so recent material is necessarily missing. Additional sources of information on chemical and radiochemical properties of the elements is the multi-volume series on the analytical chemistry of the elements and on the theory and practice of analytical chemistry (Kolthoff and Elving, 1959 to present) and the series of monographs entitled, "Analytical Chemistry of the Elements" published by the USSR Academy of Sciences (Vinogradov, 1966 to present).

A further source of detailed and tested radiochemical separation and analytical procedures is the volume of standard methods for water published by the American Society for Testing and Materials (ASTM, 1973). Although these methods were not specifically designed for high sensitivity, and as written, apply to small sample volumes, they can be scaled up for the analysis of large samples.

The remainder of this section considers separations that are particularly useful, or were specifically developed, for environmental samples, and that illustrate the application to environmental samples of the separation methods listed above.

5.4.2.2 *Ion Exchange on Solids*. Ion-exchange resins have been used very successfully in separating trace amounts of radionuclides from environmental samples, and large concentration factors can be obtained. Several advantages accrue from their use. Frequently, the separation may be performed without any preliminary concentration with carriers, and large sample volumes can consequently be handled easily without evaporation and filtration. Examples are the removal of inorganic radioiodide from milk (or water) by passing it through a column of an organic anion-exchange resin, such as Dowex 1 or 2 (Boni, 1963), or of silver chloride (Fairman and Sedlet, 1966); the removal of ^{137}Cs from water and milk samples by inorganic exchangers such as ammonium phosphomolybdate (Harley, 1972) and potassium cobalt ferrocyanide (Boni, 1966); and the tandem separation of ^{214}Bi, ^{218}Po, and ^{214}Pb from rainwater by passing it in turn through an anion-exchange resin (Dowex 1) to separate lead and polonium and a cation-exchange resin (Dowex 50) to separate bismuth (Thomas, 1973). Ammonium phosphomolybdate will also strongly absorb neptunium, uranium, and thorium from solution, and thus may be useful for actinide element separations (Ganzerli-Valentini *et al.*, 1971). A simple method for removing a large fraction of many fission products from water solution consists of suspending a "tea-bag" containing a mixed cation-anion exchange resin in the solution and stirring (Eichholz and Galli, 1970). The amount of resin needed, about 5–10 g, depends on the total ion content of the water. Folsom and Sreekumaran (1970) described the standard IAEA method for ^{137}Cs in seawater.

A modified ion-exchange resin has been briefly described by Watari *et al.* (1970), who incorporated insoluble metal salts onto standard organic ion-exchange resins to produce some specific exchangers. Two examples are nickel ferrocyanide on IRA-904 resin, which exchanges cesium (as does nickel ferrocyanide alone) and nickel ferrocyanide and calcium phosphate on IRA-904, which separates both cesium and strontium. Some information on 25 of these resins was described by Hendrickson and Riel (1975) and they appear quite promising.

Very ingenious devices and techniques have been used to separate traces of radionuclides from tens of thousands of liters of sea and fresh water to obtain sufficient activity for measurement. The methods use both ion-exchange and adsorption. Two cosmic-ray produced radionuclides, ^{32}Si and ^{7}Be, and other trace elements have been separated from samples of seawater as large as $\sim 4 \times 10^5$ l by *in situ* extraction on ferric hydroxide-loaded natural sponges (Somayajulu *et al.*, 1973; Lal *et al.*, 1964). The sponges were soaked in ferric chloride and ammonium hydroxide solution to load the sponge with the compound, and the sponges were passed through seawater for 10 to 12 hours to absorb the dissolved silicon on the ferric hydroxide. The silicon was chemically separated from the sponges, and since ^{32}Si is a low-energy beta emitter and difficult to count, its decay product, ^{32}P, was allowed to grow into secular equilibrium and then separated and counted. Very low ^{32}Si concentrations, ranging from 0.5–70 pCi/10^6 kg of seawater, were measured in this way.

A variety of very sensitive methods for the collection, separation, and measurement of cosmic-ray produced radionuclides, other natural activities, and fission products, in ocean and fresh water, air, and precipitation have been developed at the University of Washington and the Battelle Pacific Northwest Laboratories. Very large samples can be processed by these techniques and the separated radionuclides, if they emit gamma rays, are counted in very sensitive, low-background gamma-ray spectrometers (Section 6.2.3). Two examples will be described briefly. Complete details can be found in their annual reports, and the more recent one (Nielsen *et al.*, 1973), for 1972, lists all the previous reports in the series.

Perkins and Rancitelli (1971) and Silker *et al.* (1971) have reported on the construction and use of a special water sampler that will move water through a series of filter and adsorption beds 0.64 to 2.54 cm deep and 29 cm in diameter at flow rates up to 150 ml min^{-1} cm^{-2}. Retention decreases significantly at this rate, but up to 50 ml min^{-1} cm^{-2} (corresponding to a total volume flow rate of 31 l min^{-1}), retention remained constant. The principal adsorbant used in this sampler was chromatographic-grade aluminum oxide. It was found to be efficient for a number of radionuclides (^{7}Be, ^{46}Sc, ^{144}Ce, and others). Treatment with various additives improved the efficiency for some nuclides. Impregnation of the aluminum oxide with stannous chloride to reduce Cr(VI) to Cr(III) made the collection of ^{51}Cr nearly quantitative. Similar improvement in the removal of ^{226}Ra was obtained by impregnating the aluminum oxide with barium sulfate. Cation and anion exchange organic resin beds were used to remove ^{24}Na, ^{38}Cl, and ^{39}Cl from rainwater. Potassium cobalt ferrocyanide was used to remove ^{137}Cs.

The same technique has been used for the rapid separation of ^{38}Cl, ^{39}Cl, ^{38}S, ^{24}Na, ^{214}Bi, ^{214}Pb, and fission products from large volumes of rainwater (Perkins, 1969; Thomas *et al.*, 1971). Fast separation is needed because the half-lives of many of the nuclides are quite short—about 30 minutes—and large volumes (1000 liters) must be processed because the concentrations are low. The sampler can process water, at flow rates of 20–30 l min^{-1}, pumped, in series, through a particulate filter assembly, a thick bed of aluminum oxide, and beds of cation and anion exchange resins. The cation resin removes ^{24}Na, the anion resin removes ^{214}Pb and ^{214}Bi. The fission products and other cosmogenic nuclides distribute according to their chemical and physical forms.

Advantage has been taken of the ability of many metal ions to form negatively-charged complexes and thus to separate on anion exchange resins (Kraus and Nelson, 1957). Many plutonium separations are based on the fact that tetravalent plutonium forms a nitrate complex in strong (8N) nitric solution that will absorb, together with thorium, on the nitrate form of a resin. Most other metals, including iron, uranium, and the trivalent actinides are not absorbed. Since plutonium also forms a chloride complex, the thorium is removed with strong hydrochloric acid, leaving only plutonium (plus any neptunium) on the resin. The same separation can be done from hydrochloric-hydrobromic acid solution (Larsen and Oldham, 1975). Plutonium (IV) and (VI), together with thorium and uranium (VI) are absorbed on an anion exchange column. Thorium is removed with strong hydrochloric acid, plutonium, through its reduction to the trivalent state, with hydrobromic acid, while uranium remains on the column. As iron (III) is also absorbed, a preliminary separation from this element may be necessary.

Anion exchange resins can be used for separations of trace concentrations of cations based on the solubility of their compounds. When a neutral solution of radiocesium and radiostrontium is passed through such a resin in the carbonate form, the strontium will remain on the column as the carbonate. This separation of the cesium depends on the relative solubilities of the compounds, although the concentrations are too low for the solubility products to be exceeded.

A recent application of this technique is the removal of stable silver from seawater by the dithizone form of an anion resin. The silver was subsequently determined by neutron activation analysis (Robertson,

1973). Separations, similar in principle but using cation exchange separation at trace concentrations, are also possible.

The insolubility of the fluorides of the tri- and tetravalent actinides and lanthanides has been used to separate americium (VI) from curium (III) on a column of calcium fluoride powder. Curium is retained on the column while americium is eluted with 0.1M HNO_3 solution (Holcomb, 1964). Except for the fact that the ion-exchange material is in the form of a column, allowing large volumes of solution to come in contact with a small amount of solid, this type of separation has been referred to as "carrying on pre-formed precipitates" in older texts (Wahl and Bonner, 1951). Precipitation of the compound in solution also effects the same separations, and insoluble fluorides have been used for "carrying" or coprecipitating the actinides since their discovery. Separation by coprecipitation through *in situ* formation of the precipitate is generally faster—or more complete in less time—because more crystal lattice sites are available for the trace nuclide to be incorporated in the carrier compound.

5.4.2.3 *Coprecipitation*. Coprecipitation is probably the most common separation technique used for environmental separations and is the most generally applicable, since the technique is based on chemical properties that have been well-developed over many years for ordinary analytical separations.

In general, coprecipitation separations can be performed on any element that forms an insoluble compound. Some of these are fairly specific, such as the barium sulfate system discussed below, and some are very general, such as the carrying on ferric hydroxide of metallic cations that form insoluble hydroxides and carbonates. The known precipitation reactions of elements with stable isotopes are widely used for separating their radioactive isotopes, and are carried out in the standard manner after adding carrier. However, the accidental carrying of traces of elements under unexpected conditions must also be considered in devising separation schemes. Two different examples illustrate this problem. The low-energy beta emitter, ^{63}Ni, is commonly separated by adding inactive nickel carrier and precipitating nickel dimethylglyoxime, a standard analytical procedure. A preliminary purification of contaminating radionuclides is accomplished by precipitating ferric hydroxide with ammonia. Under these conditions, nickel should remain in solution as the complex ion, $Ni(NH_3)_4^{++}$. However, unless the solution contains a soluble electrolyte, such as 0.5-1M NH_4Cl or $NaNO_3$, as much as 50 percent of trace nickel will carry on the ferric hydroxide. Bondarenko *et al*., (1971) have found that ^{95}Zr-^{95}Nb could not be separated from ^{90}Y by repeated yttrium hydroxide precipitations in the analysis of ^{90}Sr in seawater.

The publication by Sunderman and Meinke (1957) should be consulted for additional experimental information on these types of difficulties.

One of the more interesting and useful coprecipitation methods has been devised by Sill and co-workers (Sill and Willis, 1964; Sill, 1969; Sill and Williams, 1969; Sill *et al.*, 1974) who discovered that barium sulfate, precipitated from sulfuric acid solution in the presence of potassium, will coprecipitate all of the large tri- and tetravalent cations. The technique removes all the heavy elements from radium through at least californium, the light lanthanides, lead, bismuth, and (less efficiently) polonium. The oxygenated higher-valent ions, as uranyl (UO_2^{++}), must be reduced to the tetravalent state to be carried. As a result, the elements from actinium through californium can also be separated from each other (the trivalent actinides as a group) by a combination of barium sulfate reprecipitations, after adjusting their oxidation states, and liquid-liquid extractions. The barium sulfate is dissolved in sulfuric acid-potassium hydrogen sulfate or aluminum nitrate solution for these subsequent separations. The barium sulfate precipitate itself can be counted for total alpha activity, or the barium sulfate can be separated from the actinides by dissolving it in diethylene triamine pentaacetic acid (DTPA) and reprecipitating the barium sulfate at pH 5 in acetic acid solution. The actinides in the supernatant can be electrodeposited for alpha spectrometry.

The tendency of polyvalent ions to form "radiocolloids" or to hydrolyze at trace concentrations has been used to advantage in interesting and potentially very useful separations. Eakins and Gomm (1968) have found that the actinide elements absorb almost quantitatively on glass fiber paper at pH 5, and can be readily removed from the paper by strong acid. Kanapilly and Yaeger (1969) have separated rare earths from biological samples by adding yttrium carrier and filtering a colloidal yttrium phosphate compound at pH 5 (or at lower pH, if the solution became cloudy) through a 0.3-μm membrane filter. The yttrium phosphate apparently acts as a carrier for the trace rare earth radionuclides.

5.4.2.4 *Liquid-Liquid Extraction.* Liquid-liquid extraction separations have been widely used for environmental samples. Usually the extraction is used for purification of a radionuclide rather than for the initial separation. Some of the extractions are fairly specific for an element, while many solvents extract a variety of elements, and the conditions are adjusted to favor the desired radionuclide. The number and variety of these separations is so large that only some general classes will be discussed here. The references to solvent extraction separations cited earlier give numerous specific examples as well as basic and theoretical information. Extraction with high molecular weight amines (Moore, 1960) is a versatile separation method, and functions much in the same manner as the organic anion exchange resins. The amines can, therefore, separate the usual anions as well as metallic ions that form negatively-charged anionic complexes. Examples are the extraction of technetium as TcO_4^- by tri-isooctylamine (Golchert and Sedlet, 1969) and the extraction of plutonium (IV) as the anionic nitrate complex by Aliquat 336, a methyltrioctyl and decyl amine (Sill *et al.*, 1974) and by tri-isooctylamine (Sakanoue and Tsuji, 1971).

Among the chelating agents used as extractants are dimethylglyoxime in chloroform for ^{63}Ni (Krieger and Gold, 1973), diethylammonium dithiocarbamate in chloroform for ^{210}Pb and ^{210}Bi (Sill and Willis, 1965), and ammonium pyrrolidine dithiocarbamate for ^{210}Po and ^{210}Pb (Shannon and Orren, 1970).

The β-diketone, thenoyltrifluoroacetone (TTA) extracts a large number of cations into organic solvents by forming water-insoluble chelates. The distribution coefficient is strongly dependent on activity, and many separations of metal ions can be obtained by proper pH adjustment. A recent example of its use was in the separation of plutonium and neptunium from pitchblende (Myers and Lindner, 1971). The use of TTA has been largely supplanted by organic phosphate esters, primarily because cations will extract into the latter compounds at higher acidities than into TTA. Consequently, the preparation of aqueous solutions for the extraction is simpler and less exacting. For example, while trivalent ions will extract into a TTA-xylene solution at a pH no less than about 5, extraction into di(2-ethylhexyl)phosphoric acid in xylene occurs from 0.1N mineral acid. Tributylphosphate, which was originally used for separations of uranium from fission products because of its good resistance to radiation damage, is frequently used for the separation of ^{90}Y from ^{90}Sr (Douglas, 1967) after the strontium fraction has been separated from environmental samples, and for the separation of tri- and tetravalent ions (Menon *et al.*, 1963). Numerous other applications have been found for the organic phosphates (Peppard *et al.*, 1963; Peppard, 1971).

A very promising separation system utilizes an organic phosphate ester, usually di(2-ethylhexyl)phosphoric acid (HDEHP) absorbed on an inert solid such as Celite. In column operation, chromatographic separations are obtained through numerous liquid-liquid extraction plates, so the technique is referred to as extraction chromatography. Two examples of this technique are given for a particularly difficult problem, separation of ions of like charge by liquid-liquid extraction with HDEHP. Horwitz and Bloomquist (1973) report complete, or nearly complete, separation of adjacent pairs of tri-positive actinides, and extremely good separation of actinides differing by two atomic numbers, on columns of HDEHP absorbed on Celite. Americium has been separated from other transuranic elements on a HDEHP column by oxidizing it to the pentavalent state with persulfate ion. In this oxidation state it is not absorbed on the column from 0.01–0.05M HNO_3, while the actinide (III, IV) and lanthanide (III) ions are retained on the column. If the sample contains californium and curium, the latter can be eluted from the column with 0.3M HNO_3, and the former with 4M HNO_3 to obtain additional separations (Moore, 1968). This type of separation should receive more attention for analyses of environmental materials, as should the synergistic effects possible in solvent extraction (Irving and Edgington, 1960).

A useful technique combines solvent extraction of a radionuclide, frequently into organic phosphates, with liquid scintillation counting. The extract is added to an organic-based liquid scintillation solution and counted directly. Organic phosphates are used because they do not quench the counting system as severely as other extractants, ethers, for example. Although any type of emitter can be counted, the method is particularly useful for low-energy beta emitters (e.g., ^{147}Pm, ^{35}S, ^{210}Pb) and electron capture nuclides (e.g., ^{55}Fe). A recent example of its use is the determination of uranium in water by extraction into tributylphosphate, and counting of the extract in a liquid scintillation system (Grobushina *et al.*, 1972). One of the difficulties with this counting method is that the backgrounds of commercial liquid scintillation counters are much higher than the counters designed to count solid samples. This places severe sensitivity limitations on this technique. Improvement in backgrounds may be expected in the future, as indicated by the performance of recently-reported liquid scintillation counters that are much more sensitive for tritium than commercial counters (Noakes *et al.*, 1973; McDowell *et al.*, 1974).

The recently discovered hepta- and octavalent states of neptunium and plutonium that can be prepared in strong alkaline solution offer new possibilities for the separation of these elements from most other metals. The neptunium (VII) ion, NpO_5^{-3}, has been efficiently extracted from sodium hydroxide solution by the β-diketone, dibenzoylmethane (Novikov *et al.*, 1972).

A few other separation techniques will be mentioned briefly. A collection of distillation or volatility separations is given by DeVoe (1962). Included are methods for many elements, including the rare gases that are of particular current interest. Electrodialysis has been used by Owers (1959) to concentrate radiostrontium and radiocesium from 100-l samples of water. Concentration factors of about 95 for cesium and 30 for strontium were obtained in 5.5 h of dialysis. The foam separation of radiocesium by insoluble heavy metal ferro- and ferricyanides has been reported by Pushkarev *et al.* (1960). Foaming was produced by adding gelatin. Apparently this technique has not yet been applied to environmental samples.

5.4.3 *Preparation of Separated Radionuclides for Counting*

The preparation of the separated activity in a reproducible form for accurate measurement requires some care and attention. The technique of direct liquid scintillation counting of the extracted activity described in Section 5.4.2 has much to recommend it if adequate sensitivity is obtained. The sensitivity will be a function of the original sample size, counter background, and counting efficiency. Liquid scintillation counting is subject to numerous quenching and enhancing effects, and care is required to obtain the same counting efficiency between samples and between sample and standard, or otherwise to measure the counting efficiency for each sample. Color quenching in the liquid scintillation counting of ^{55}Fe is a good example of this problem. The amount of quenching depends on the stable iron concentration, which must be determined for each sample, and a correction made from a previously-measured curve of quenching *vs* concentration. This quenching is eliminated in a method devised by Eakins and Brown (1966), who convert the iron to a colorless phosphate complex. Two of the low-energy beta emitters often measured by liquid scintillation counting, ^{35}S and ^{14}C, form acidic oxides that can be absorbed in organic bases (Bosshart and Young, 1972; Gupta, 1966). Commercial manufacturers of liquid scintillation counters are a good source of information on the preparation of samples for liquid scintillation counting.

The standard method of preparing a powder or precipitate for presentation to a counter by filtration or centrifugation onto a flat filter paper or porous plate, or by evaporation of a slurry on a planchet, is described in the texts on radiochemistry. A smooth, uniformly thick deposit is required for accurate and reproducible measurements of short range radiations – alpha and beta particles. In this respect the article by Libby (1957) on the effect of energy, sample smoothness, and backscattering on the accuracy of beta counting should be consulted. A centrifugation method for obtaining a very uniform deposit of barium sulfate for alpha counting is described by

Sill (1969). Scintillation counting of alpha particles by mixing the radionuclide immediately with silver-activated zinc sulfide fluorescent powder is a useful method of obtaining very high alpha counting efficiencies and, at the same time, avoiding many of the uniformity and thickness problems of ordinary alpha counting. The zinc sulfide powder will remove a number of alpha emitters (polonium, plutonium, others) from solution by adsorption or ion exchange when the sulfide is mixed in solution at the proper acidity, and the powder is then separated and placed on the face of a multiplier phototube for counting (Rosholt, 1957; Morken, 1959).

Deposits suitable for alpha spectrometry must be thin and uniform. Additionally, if solid state silicon diodes are used for detection, the area of the deposit must be small, 300 mm^2 or less, because the detector areas are usually 400–500 mm^2. Frisch-type gridded ion chambers can count larger area samples with good efficiency (but poorer resolution), but they are not generally available.

Electrodeposition of alpha emitters is an excellent method for preparing such deposits, and is preferred by most investigators. The method is simple, rapid, and quantitative under the proper conditions. In some respects, the technique is partially an art, and each laboratory worker must test the method for himself. Deposition of the actinide elements on the cathode of an electrolysis cell is carried out from a slightly acid solution (pH 1–2) at one to a few amperes, from 15 minutes to several hours, depending on the system employed. Before the electrolysis is stopped, the solution must be neutralized with ammonium hydroxide to prevent dissolution of the radionuclide on the cathode. It is believed that the metallic ion deposits on the cathode as a hydrous oxide [sometimes after reduction to a lower oxidation state, as with uranium (VI)] as the acidity in the vicinity of the cathode is reduced by electrolysis of hydrogen ion. Successful electrodepositions have been obtained on platinum and steel cathodes from a variety of electrolytes: NH_4Cl-HCl (Mitchell, 1960); NH_4Cl-$(NH_4)_2C_2O_4$ (Puphal and Olsen, 1972; Baines, 1963); $(NH_4)_2SO_4$-H_2SO_4 (Talvitie, 1972); and NH_4NO_3-HNO_3 (Bärring, 1966). The electrodeposition of actinides is carried out from relatively pure solution, after considerable chemical purification and separation from environmental samples. However, microgram amounts of various cations generally remain and can cause interferences, either by depositing on the cathode and giving a source too thick for good alpha particle resolution or by precipitating, and carrying with it the actinide ions, if the solution is made basic during the pH adjustment (redissolution of the basic compounds may be slow) (Puphal and Olsen, 1972). Various complexing agents have been used to eliminate these interferences, including oxalate ion, fluoride ion, and chelating agents such as DTPA.

Many radionuclides other than the actinides can be electrodeposited for counting and this is advantageous if they decay by electron capture or emit only low-energy beta particles. The deposited source is uniform and smooth, and if the metal can be deposited, the source has the highest specific activity obtainable. Among such nuclides are ^{99}Tc, ^{63}Ni, ^{55}Fe, and ^{57}Co. The latter three metals can be electroplated from solution with stable carrier if desired. If routine electrodepositions are performed, disposable electrolysis cells are quite useful (Harley, 1972).

Although electrodeposition is the preferred method for obtaining a good source for alpha spectrometry, and will generally give the best resolution and the most information, it is not always essential. Adequate resolution and quantitative results for plutonium have been reported by Butler *et al.* (1971), who prepared the source by coprecipitating the plutonium with 0.1 mg lanthanum(III) as the fluoride, and filtering the precipitate on membrane paper.

A potentially useful preparation technique for beta emitters that also includes the possibility of chemical separations was reported by Dobosz (1973) in a radiochemical iodine procedure. Iodide ion was separated and counted on an anion exchange resin film (22-mm diameter by 2-mm thick) prepared by spreading 0.5 g of the resin saturated with styrene on a filter paper and polymerizing the styrene with ultraviolet light. The resulting film is reported to be permeable to liquids and remain mechanically intact during filtration and counting.

5.4.4 *Examples of Radiochemical Analytical Procedures*

As a summary for this section, Table 5-6 gives some typical methods that have been used successfully for environmental samples. The table is not exhaustive, but is intended to be representative of the types of radiochemical procedures currently in use for environmental samples. Purely instrumental methods that do not require significant sample handling or modification for their use, notably gamma-ray spectrometry, are not included. The variety of methods given for the same analysis should leave the impression that there is generally more than one procedure that is satisfactory for a particular analysis. While some techniques may at a given time be the method of choice, progress and improvements will cease unless some investigators are willing to use or study others. The detection limit given in the table is

TABLE 5-6—*Summary of radiochemical procedures for environmental samples*

Analyte	Sample	Preparation; Dissolution	Chemical Separation	Method of Measurement	Detection Limit[a]	Reference
^{7}Be ^{24}Na ^{38}Cl ^{39}Cl	Precipitation, air	Filter paper—leach with acidic carrier solution	Cl separated as AgCl; ^{24}Na measured in residue; other radionuclides removed with cation resin; Be separated as hydroxide	Two parameter gamma-ray spectrometry	10 fCi l^{-1} (precipitation) 0.05 fCi m^{-3} (air)	Thomas *et al.* (1969)
^{14}C	Food, air	Combust, to CO_2	CO_2 purified by expansion, condensation, drying agents	Internal gas counter, anticoincidence shielded	5 pCi g^{-1} (carbon)	Drobinski *et al.* (1965)
^{14}C	Water	Add oxalate carrier	C oxidized to CO_2 in acid solution, $CaCO_3$ collected	Beta counter (gas flow or liquid scintillation)	10–100 pCi (depending on C content)	Krieger and Gold (1973)
^{24}Na	Precipitation	Add Na carrier	Na separated by ion exchange, hydroxides scavenged, Na measured in filtrate	β-γ coincidence counter	0.02 pCi l^{-1}	Yasyulenis *et al.* (1973)
^{32}P	Sea water	Add PO_4^{-3} carrier	P oxidized with $HClO_4$, NH_4 phosphomolybdate extracted into isoamyl alcohol; precipitated as $MgNH_4PO_4$	Beta counter	1 pCi l^{-1}	Flynn and Meehan (1973)
^{55}Fe	Various		$Fe(OH)_3$ precipitated, dissolved, extracted, and added to scintillation mixture; or hydroxide dissolved in H_3PO_4-NH_4Cl and colorless complex counted	Liquid scintillation counter	0.5 pCi g^{-1} (iron)	Cosolito *et al.* (1968); Eakins and Brown (1966)
^{131}I	Milk	Add HCHO and I carrier	I separated on anion-exchange resin, purified by extraction; PdI_2 precipitated	Beta counter	0.25 pCi l^{-1}	USAEC (1973b)
^{137}Cs	Milk, bone, vegetation ash, soil, water	Fuse—Na_2CO_3 or leach—HNO_3	Ca precipitated; Cs separated on NH_4 phosphomolybdate, purified by ion exchange, precipitated as Cs_2PtCl_6	Beta counter	1 pCi per sample	Harley (1972)

TABLE 5-6—*Continued*

Analyte	Sample	Preparation: Dissolution	Chemical Separation	Method of Measurement	Detection Limit[a]	Reference
^{137}Cs	Milk, water	—	Ion-exchange on K cobaltihexacyanoferrate	Gamma-ray spectrometry	0.01 to 10 pCi l^{-1} (varies with volume)	Boni (1966)
^{137}Cs	Sea water	—	Cs separated on NH_4 phosphomolybdate, silica gel mixture; SiO_2 dissolved with HF	Gamma-ray spectrometry	0.1 pCi l^{-1}	Terada *et al.* (1970)
^{141}Ce ^{144}Ce	Precipitation	Add Ce carrier	Ce separated by oxidation-reduction and solvent extraction cycles	Gamma-ray spectrometry with Ge(Li) detector	1 pCi per sample	Menon *et al.* (1963); Sherwood *et al.* (1973)
^{210}Pb	Bone	Ignite—dissolve in HBr	Pb extracted into quaternary amine; $PbSO_4$ precipitated and counted	Beta counter (^{210}Bi counted)	1 pCi per sample	Petrow and Cover (1965)
^{210}Pb	Water, soil, ores	Treat with acids, fuse—KF and $Na_2S_2O_7$, dissolve—HCl	Pb and Bi extracted into diethyl dithiocarbamate; ^{210}Bi separated from Pb by extraction in dithizone	Beta counter (^{210}Bi counted)	1 pCi per sample	Sill and Willis (1965)
^{210}Pb ^{210}Po	Bone, tissue	Dissolve in mineral acids	^{210}Po deposited on Ag from HCl, ^{210}Pb determined by deposition of ^{210}Po after ingrowth	Alpha counter	0.02 pCi per sample	Holtzman (1963)
^{210}Pb ^{210}Po	Sea water	Add carriers	Pb and Bi carriers precipitated with $CaCO_3$, separated by hydroxide precipitation, ^{210}Po deposited on Ag for counting; ^{210}Pb determined by deposition of ^{210}Po after ingrowth	Alpha counter	0.1 pCi per sample	Nozaki and Tsunagai (1973)

TABLE 5-6—*Continued*

Analyte	Sample	Preparation; Dissolution	Chemical Separation	Method of Measurement	Detection Limit[a]	Reference
^{226}Ra ^{228}Ra	Soil vegetation	Plant—dissolve in acids; soil—exchange with NH_4 acetate	^{225}Ra, ^{133}Ba tracers and Ba carrier added; Ra precipitated with $BaSO_4$, separated from Ba by ion exchange; electrodeposited for counting	Alpha spectrometry	1 fCi g^{-1}	Smith and Mercer (1970)
^{228}Ra ^{228}Th	Vegetation, food, soil, bone	Ash—dissolve mineral acids; soil—fuse Na_2CO_3	Th extracted by ethylhexyl phosphoric acid, ^{224}Ra daughter removed with HNO_3 after ingrowth and counted; Ra precipitated with $PbSO_4$ purified as $PbBr_2$, ^{228}Ac daughter extracted and counted	Alpha and beta counters	1 pCi Ra, 20 fCi Th per sample	Petrow *et al.* (1964)
^{232}Th	Bone	Ignite at 600°C	Irradiate with neutrons, dissolve in HCL, separate ^{233}Pa by anion exchange	Gamma-ray spectrometry of ^{233}Pa	0.1 ng g^{-1}	Lucas *et al.* (1970)
Th	Various	Ash, dissolve with acids and	Coprecipitate with $BaSO_4$, dissolve in DTPA	Fluorescence of Th-morin complex	10 pg ml^{-1} 10 ng g^{-1}	Sill and Willis (1964)
^{90}Sr	Bone	Ignite, dissolve in HCl	^{90}Y extracted into ethylhexyl phosphoric acid; impurities removed by amine extraction	Beta counter	1 pCi per sample	Petrow (1965)
^{90}Sr	Food vegetation, tissue milk, soil, water	Evaporate, ignite, fuse with Na_2CO_3; Soil—fuse with Na_2CO_3 or leach with NaOH, HCl; ^{85}Sr and Sr carrier added	Sr separated, purified by nitrate precipitations, ^{90}Y precipitated and counted	Beta counter	0.5 pCi per sample	Harley (1972)

TABLE 5-6—*Continued*

Analyte	Sample	Preparation; Dissolution	Chemical Separation	Method of Measurement	Detection Limit[a]	Reference
^{90}Sr ^{137}Cs	Sea water	Acidify—HCl; added carriers	Cs removed by NH_4 phosphomolybdate, purified, precipitated as Cs_2PtCl_6; Sr precipitated, ^{90}Y daughter separated and counted	Beta counter	0.5 pCi per sample	Sutton and Kelly (1968)
^{99}Tc	Water	—	Tc purified by precipitation, solvent extraction; electrodeposited for counting	Beta counter	0.5 pCi l^{-1}	Golchert and Sedlet (1969)
^{108m}Ag ^{110m}Ag	Biological Samples	Grind in blender with water	Slurry treated with NaOH-AgCN solution; Ag electroplated on Pt, dissolved, AgCl precipitated	Gamma-ray spectrometry (4π coincidence counter)	1 pCi per sample	Hodge and Folsom (1970)
^{129}I	Milk, Water	—	Milk—I first separated by anion exchange. Milk and water—I extracted into CCl_4 and toluene; and decolorized with UV light in presence of 2-methyl-l-butane	Liquid scintillation counter	0.3 pCi l^{-1}	Gabay *et al.* (1974)
Th	Various	Dissolve as necessary	Coprecipitated with Ca oxalate, purified by ion-exchange	Colorimetric, with thorin or other dyes	1 μg	Harley (1972)
th-Cf	Soil	Treat with acids, fuse—KF, $Na_2S_2O_7$, dissolve—HCl	Actinides coprecipitated with $BaSO_4$, separated by solvent extraction after adjusting oxidation states, electrodeposited for counting	Alpha spectrometry	1 fCi g^{-1}	Sill *et al.* (1974)
U	Water	—	Extraction into methyl isobutyl ketone	Fluorescence of U in fused NaF	5 ng l^{-1}	ASTM (1973)

TABLE 5-6—*Continued*

Analyte	Sample	Preparation; Dissolution	Chemical Separation	Method of Measurement	Detection Limit[a]	Reference
^{237}Np ^{239}Pu	Pitchblende	Dissolve in mineral acids	Np(IV) extracted into trifluoroacetylacetone, purified by anion exchange; Pu(III) coprecipitated with LaF_3 and UF_4, purified by anion exchange and electrodeposited	Np—mass spectrometry Pu—alpha spectrometry	1 fCi Pu g^{-1} (sample) 0.4 ag Np g^{-1} (sample)	Myers and Linder (1971)
Pu	Soil, air particulates	Soil—ash, extract with hot HCl; particulates—ash, dissolve—mineral acids and fusion	Pu separated and purified by cation and anion exchange; electrodeposited for counting	Alpha spectrometry	1 fCi g^{-1}	deBortoli (1967)
Pu	Soil	Fuse—Na_2CO_3, dissolve—HNO_3	Phosphates precipitated and Pu purified by anion exchange; electrodeposited for counting	Alpha spectrometry	1 fCi g^{-1}	Harley (1972)
Pu	Soil	Leach with mineral acids	Pu purified by anion exchange; electrodeposited for counting	Alpha spectrometry	1 fCi g^{-1}	Chu (1971); Harley (1972)
Pu	Water, air, soil, food, vegetation	Treat with mineral acids	Pu purified by anion exchange; electrodeposited for counting	Alpha spectrometry	1 fCi g^{-1}	Wessman *et al.* (1971)

[a] Limits are as reported in, or inferred from, references. Note that these limits are expressed in several units: weight, weight/volume or weight, activity, activity/sample, activity/volume or weight, and activity/weight element (i.e., specific activity).

that specified by the author, or where no limit is given in the original publication, the limit has been calculated from the sample size, counter background, and other parameters as indicated in Section 6.1.2.

5.5 Analysis of Stable Elements

The measurement of certain stable elements at the concentrations that occur in environmental materials is of considerable importance in the study of environmental radioactivity. In studying the environmental behavior of radionuclides—their movement and transport through the ecosystem and their pathways from source to man—the concentrations of stable isotopes of the radionuclides and of elements chemically similar to the radionuclides are often needed to interpret the results of radionuclide measurements and consequences of radioactivity in the environment. For manmade and cosmogenic elements that enter the environment, the rate at which they mix with the main body of the stable elements is important for dose assessment, for the study of biogeochemical processes, and for the study of geological behavior. An example of the importance of stable element analysis is cited by Perkins and Ranticelli (1971), who report that the specific activity of ^{55}Fe is higher in fish than in ocean water. This indicates that the ^{55}Fe produced in nuclear tests is in a chemical form more readily available to the fish than is the natural iron in the ocean. Consequently, the dilution effect on dose that would be expected by complete isotopic exchange in the ocean is not occurring.

There is a purely analytical reason for stable element analysis in certain instances. If the yield of a radiochemical analysis is determined by the fraction of stable element carrier recovered, and the sample naturally contains sufficient quantities of the element to affect this result, then a separate measurement of the natural content must be made to obtain good accuracy. This situation exists, for example, in the determination of radiostrontium and radiobarium in many soils and waters and of radioiodine in milk.

The measurement of ^{137}Cs and ^{90}Sr in the environment should be accompanied by the measurement of the chemically similar potassium and calcium, respectively. The ratios of ^{137}Cs/K (stable) and ^{90}Sr/Ca (stable) are commonly used to express the activity of these nuclides in environmental samples, particularly in milk. Since potassium and calcium are present in natural minerals in high concentration, conventional chemical methods can be used. However, it is usually faster, more economical, and much simpler to perform these analyses by instrumental techniques. The laboratory manuals referred to earlier give procedures for potassium and calcium in environmental samples.

The most widely used and recommended technique for stable element analysis is atomic absorption spectrometry. It is rapid, has adequate accuracy, can be used for most elements of interest, and good equipment is available commercially at reasonable cost compared to other instruments. The manual by Harley (1972) gives methods for analyzing a number of trace elements by this method. For additional information the literature on atomic absorption spectrometry should be consulted.

Neutron and charged particle activation analysis is also an excellent, sensitive and widely applicable method for analysis of trace elements (Perkins and Ranticelli, 1971; Edgington and Lucas, 1970). It is more expensive than atomic absorption, not only because a source of neutrons or particles is needed, but because sophisticated counting equipment is also required. However, laboratories engaged in environmental radioactivity analyses can undoubtedly use the same counting equipment for activation analysis.

Other methods of trace analysis can also be used, and the choice is frequently determined by availability of equipment, the specific experience of the laboratory staff, and prejudice.

6. Laboratory Measurements

6.1 General Principles

6.1.1 *Introduction*

Problems including relatively high cost in the chemical separation of radionuclides can often be circumvented by relying on nuclear instrumentation for determining the amount of a particular nuclide present in a sample. Instrumental methods have their own problems, however, and several factors bear on the selection of appropriate measurement methods. The dosimetric importance of a radionuclide is not necessarily proportional to its concentration in a mixture and analysis for the significant radionuclide generally must be accomplished in the presence of greater concentrations of other nuclides. These interferences can be reduced by proper choice of measurement method.

As a result of public interest and concern, some analyses are required to be more sensitive than in the past (FRC, 1960; NCRP, 1971). As a result, larger samples must often be collected to provide sufficient activity to be counted in a reasonable length of time. Attempts to determine low concentrations also mean that natural and fallout nuclide interferences are more severe and tax the capabilities of the methods. Efforts to guard against transfer of contamination among samples must receive more attention and this requirement sometimes causes instrumental analyses to be favored over chemical techniques because of the minimum sample handling required by the former. All of these factors lead to increased cost of analysis and indicate the need for careful selection of analytical techniques.

As waste handling and treatment systems are improved, effluents will not only contain less radioactivity but may also have different compositions. The resulting variation in radionuclide composition places an additional burden on analytical techniques developed for compositions encountered in the past.

Economies effected through reduced sample numbers, simplified radiochemical analyses, or undue use of "total beta", "total alpha", or "total gamma" techniques may result in serious radionuclide identification and quantification errors.

The considerable expertise and instrumentation required for complex analyses should be acquired by special low-level laboratories, rather than by every laboratory. There should, therefore, be increased use of direct instrumental analyses to avoid many of the problems associated with sample preparation, radiochemical separation, and radionuclide measurement. The lower attenuation of gamma rays (compared with betas and alphas) by large samples favors, in many cases, gamma-ray spectrometry.

6.1.2 Errors and Detection Limits

Errors in radioactivity measurements are attributable to (a) the random nature of radioactive decay, (b) lack of instrument sensitivity or stability, (c) inaccuracies in the calibration radionuclide standards and the calibration procedures, (d) uncertainties in the decay properties of the radionuclides, and (e) other errors such as those due to variations in source positioning or geometry. Because of the stability of many modern instruments, the errors resulting from their use are essentially those associated with the random nature of the decay process itself. Such errors can be estimated with the relatively simple statistical treatment, but estimation of total error requires many repeated measurements and consideration of possible sources of systematic errors, such as may be due to c, d, and e.

Most counting instruments are calibrated by reference to standard sources having known error limits, which must be incorporated into the overall error. Refinements in knowledge of the radionuclide decay schemes may involve significant changes in the branching fractions and emission energies and change the uncertainties in the amounts of radionuclides determined. Errors associated with these latter uncertainties are usually, but not always, small.

Detection limit requirements for analytical procedures and measurements are not easily determined. Although limiting doses and concentrations recommended by advisory groups indicate such limits, regulatory bodies and public concern are now requiring operation below these standards. In this section, procedures and apparatus are described which may be capable of measuring radionuclides at a level sufficiently low to meet today's conservative standards. These techniques require large samples, high efficiency detectors, low background shielding, and long counting times. At such low counting rates, counting statistics become extremely important.

The process of determining the errors and hence of estimating the reliability of the measurements depends on a substantial understanding of statistical methods, as well as experience in the interpretation of low-level data. There is not space in this report for a discussion of the techniques and methods involved, so for introductory treatment and references on the statistical handling of counting data, the reader is referred to Section 7, Statistics, in a forthcoming NCRP report, (NCRP, 1977). This reference also reviews the subject of detection limits and their determination. Additional references to this subject are Currie (1968); Altshuler and Pasternack (1963); and ICRU (1972). As a practical measure, many workers define the detection limit as that sample counting rate which is equal to two or three times the standard deviation of this rate.

6.2 Counting Instruments

6.2.1 *Alpha Activity*

The quantitative measurement of alpha particles depends on difficult chemical separation procedures to isolate specific elements in a relatively pure and low mass form, because of the very short ranges of these particles. Some care is needed to prevent contamination of the detector during operation by the daughter products of radon in the air.

Although it is possible to count alpha-emitting radionuclides in thick sources, counting efficiency is reduced and thin deposits are more suitable for low-level samples. For instance, only about 5 percent of the 4.5-MeV alpha particles present in a 4-mg cm^{-2} thick organic source will escape, whereas almost all of the alpha particles will escape a 0.1-mg cm^{-2} thick source.

Most procedures for alpha emitters require a prior knowledge of which radionuclides are present in the sample. Alpha emitters not specifically isolated may be accounted for by the use of "total alpha" counting. Such counting can be applied only where the sample is very finely ground and in a low residue form spread over a sufficiently large area.

An alpha particle spectrometric method for materials of specific activity as low as a few tenths of a picocurie per gram was developed by Hill (1961). Spectral peak widths of about 150 keV were obtained with a gridless pulse ionization chamber large enough to accept a finely ground source up to 1.5 grams spread over an aluminized cellulose acetate sheet electrode of 1.5 m^2. Use of this instrument allows all of the alpha emitter activity in many environmental samples to be determined directly or on sample ash without losses due to chemical separation and at appropriately low levels.

Unfortunately, this technique has not received wide acceptance. Osborne and Hill (1964) developed smaller and less sensitive chambers. These gridded chambers exhibit spectral peak widths as low as 50 keV and accommodate samples with source areas of 20 to 1500 cm^2.

Other types of alpha spectrometers are commonly used after chemical separations (Nielsen and Beasley, 1967). Gas flow proportional counters have a very high efficiency, especially if the source is placed directly inside the chamber volume. Counting gas is used to flush air from the chambers and a continuous low flow maintains the conditions suitable for proportional counting. Although thin (<1 mg cm^{-2}) mylar windows reduce the counting efficiencies somewhat, they reduce contamination and handling problems.

Routine alpha counting with a thin layer of activated zinc sulfide cemented to the window of a photomultiplier tube is common. Counting efficiency approaches 100 percent and the geometry approaches 2π. Gaseous alpha emitters are measured using a zinc sulfide layer on the inner surface of a transparent counting chamber (Lucas, 1957).

Surface barrier semiconductor detectors are presently the most suitable detectors for alpha particle spectrometry, especially for small area, very low mass sources. Pulse ion chambers, gas flow proportional counters, and NaI(Tl) or CsI(Tl) scintillation detectors are also used, but the surface barrier detectors have the best energy resolution. For 5-MeV alpha particles, the resolutions are about 10 keV (full width at half maximum peak height, FWHM) for the surface barrier detectors, 15 keV for pulse ion chambers, 75 keV for proportional counters, and about 100 keV for the scintillation detectors. Surface barrier detectors can most easily resolve energies in mixtures of alpha emitters, while low backgrounds of about 0.02 cm^{-2} h^{-1} for the gas flow proportional counter and about 0.05 cm^{-2} h^{-1} for a silicon detector aid the overall sensitivity.

A counting technique having a detection limit equal to or better than that of the silicon detector, but which provides no spectral data, is the nuclear track method (Schwendiman and Healy, 1958). The alpha emitter, plated on a metal disc, is placed in contact with a nuclear track emulsion. After exposure of week or longer the film is developed, the tracks are counted by means of a microscope, and activities as low as 0.03 min^{-1} determined. This technique is easily adapted to a large number of samples but has the disadvantage of requiring microscopic analysis.

Nuclear track etch techniques (Fleischer *et al.*, 1965; Becker, 1973) have been applied to environmental samples. Flesicher and Lovett (1968) describe the application of track etch techniques for measuring alpha particles and fission fragments to the analysis of collected samples. These methods are capable of very high sensitivity and also have promise for *in situ* measurements as indicated in Section 4.4.1.

6.2.2 *Beta Activity*

Some important radionuclides either emit no gamma rays or emit gamma rays of insufficient energy or intensity for practical measurements. The most important include ^{3}H (0.0185 MeV), ^{14}C (0.156 MeV), ^{32}P (1.71 MeV), ^{35}S (0.167 MeV), ^{45}Ca (0.252 MeV), ^{85}Kr (0.67 MeV), ^{90}Sr (0.546 MeV), ^{90}Y (2.27 MeV), and possibly ^{147}Pm (0.225 MeV).

Robinson (1967) and many others have reviewed the factors involved in beta counting measurements. Commonly used detectors have sufficient sensitivities for measuring the beta-ray energies at the low levels required for environmental monitoring, when properly shielded. Beta detectors can be shielded with a few centimeters of lead or mercury and, for low-level counting, plastic scintillation or gas filled counters can provide anti-coincidence shielding. Commercial gas discharge (Geiger-Mueller or proportional) and liquid scintillation counters are available in forms suitable for the lowest level counting. Gas discharge counters are often more sensitive by as much as a factor of ten and are best suited for gas or solid sample counting. Liquid scintillation counting is suited for gas or liquid samples and requires simpler sample preparation.

Geiger-Mueller or proportional counters for beta counting are short squat cylinders or hemispheres about a few centimeters in diameter and a centimeter or less thick, provided with a thin window on one face. Common window thicknesses of 0.5–0.7 mg cm^{-2} allow detection of nearly 90 percent of the incident betas from low-energy emitters and have greater efficiencies for higher energy betas. Tritium betas cannot penetrate this thickness and even windows as thin as 0.1 mg cm^{-2} reduce the incident betas by about 70 percent. Therefore, the usual gaseous or liquid tritium, and samples of other gaseous beta emitters are usually measured with windowless counters. Gas discharge counters should be operated in an anticoincidence shield and the whole array shielded by lead or steel to achieve backgrounds of the order of a few tenths of a count per minute.

Liquid scintillation counters have sensitivities approaching those of gas discharge counters. Other characteristics that favor them for gas or liquid sample beta counting include the possibilities of employing photomultiplier tubes in coincidence and pulse height energy analysis. Spectrometry done with these scintillation detectors, as well as the gas proportional and semiconductor detectors, is generally restricted to the determination of the maximum energy of the beta spectrum or the measurement of conversion electrons.

The types of detectors described above are suitable for most of the environmental sample beta counting needs.

6.2.3 *Gamma- and X-Ray Spectrometry*

Photon spectrometry, the most widely used technique for the measurement of environmental radionuclides, has the ability to identify and measure individual radionuclides in mixtures, which is good to excellent depending on the detector type used, and to measure large samples with minimal sample preparation and minimal chance of sample loss or contamination. Backgrounds for gamma counters are considerably higher than for alpha and beta detectors so that extensive bulk or anticoincidence shielding is required to obtain a gamma counting system capable of low-level operation. Scintillation, gas proportional, and semiconductor detectors are all used for spectrometry.

6.2.3.1 *Scintillation Detectors*. Scintillation detectors are widely used because of their relatively low cost and high efficiency, although their energy resolution is poor. Gas proportional counters are intermediate in resolution and are generally restricted to spectrometry below 100 keV because of practical limits on size and gas pressure. Semiconductor detectors have the highest resolution and, in spite of their relatively high acquisition cost, are the detector of choice for most applications provided sufficiently large ones are available. Each significant increase in the size of semiconductor detectors is accompanied by increased usefulness.

NaI (Tl) detectors are the most widely used for gamma-ray spectrometry because of the large light output per incident photon, relatively high atomic number and density, high transparency to its own fluorescence light, which also has a relatively short decay time (0.25 μs). Many sizes are available and those of the same actual size and shape have the same photon counting efficiencies so that calibration data obtained for one detector will be nearly identical for other similar detectors (Section 4.5). Standard data and spectra for 7.6 × 7.6-cm NaI(Tl) detectors have been published by Heath (1964) and are useful for selecting proper detector sizes. Data on the more efficient and sensitive 10 × 10-cm crystals have been published (for example PHS, 1967) and indicate that this crystal provides suitable sensitivity for most analyses required for nuclear facilities surveillance.

In some cases a 3.5-l environmental sample is counted directly in a Marinelli beaker, a cylindrical container having a reentrant cavity in its bottom into which the scintillation crystal fits so that the sample nearly surrounds the crystal. Perkins (1961) described a more efficient, large cylindrical well detector having overall dimensions of 24 × 22 cm with a 7.6-cm diameter well extending 7.6 cm into the crystal. This detector has about ten times greater sensitivity for most nuclides than a 7.6 × 7.6-cm crystal for the usual 500-ml samples. For some 15 radionuclides studied, the sensitivity was sufficient to measure better than 0.1 percent of the maximum permissible concentration recommended for designated radionuclides in drinking water (NCRP, 1959). The principal advantage of this large detector is that a smaller sample size or lower sample concentration can be analyzed. Because the background counting rate increases with crystal size, an effect limiting the sensitivity, large crystals are often enclosed in a liquid, plastic, or NaI(Tl) scintillator anticoincidence shield to reduce the background counting rate (Perkins *et al.*, 1960). An anticoincidence shield also serves to reduce the recorded continuum spectrum by rejecting degraded gamma rays, making the photopeaks easier to measure, and providing greater effective sensitivities in the analysis of mixtures.

Individual radionuclides in complex mixtures cannot easily be analyzed by single crystal NaI(Tl) spectrometry, but computer methods can be used as indicated in Section 6.2.3.3 and described in a forthcoming NCRP report (NCRP, 1977). Chemical separation into either groups or individual elements should first be performed, but multiple detector and coincidence techniques often permit direct instrumental analysis. More than one half of the gamma-emitting nuclides decay through the emission of two or more gamma rays in cascade. By accepting only those counts that result from simultaneous events, analyses of individual radionuclides can be made which are impossible by single crystal spectrometry.

Special techniques for accomplishing these analyses are: (a) sum coincidence spectrometry where sum pulses are recorded that correspond to the absorption of the full energies of the two (or more) cascade gamma rays (Hoogenboom, 1958); (b) pair spectrometry employing a three crystal triple-coincidence system where only events are recorded in the main crystal that are in coincidence with the detection of each of the positron annihilation photons in the other two crystals (Johansson, 1950); and (c) multidimensional gamma-ray spectrometry. In the latter and most versatile technique, each event is stored in a multichannel analyzer or computer memory according to the energy deposited simultaneously in each crystal. This provides a great increase in selectivity, and because of background reductions of several orders of magnitude, great sensitivity improvements are obtained (Perkins, 1965).

6.2.3.2 *Proportional Counters*. Gas-filled proportional counters are effective in the range of 1–100 keV. They have some advantages over scintillation spectrometers, but are not as useful as semiconductor detectors. Gas proportional counters are advantageous because (a) absorption in the gas occurs almost wholly by the photoelectric process giving total energy peaks, (b) the energy resolution is two to three times better than that in NaI(Tl) below 100 keV, (c) the beryllium or aluminum detector window can be thinner than for NaI(Tl) so that photons down to about 1 keV can be analyzed, and (d) by suitable construction and with massive and anticoincidence shielding, backgrounds of sample chambers of a liter or more can be made as low as 10 min^{-1} over a wide energy range.

Proportional counters are large and sometimes operated at several atmospheres pressure in order to maintain a high absorption efficiency for x rays in the gas filling. The most common gas filling in gas proportional counters is 90 percent argon and 10 percent methane by volume.

Other rare gases may be used in place of argon. Helium and neon have smaller absorption coefficients; krypton and xenon have higher coefficients, but interpretation of spectra measured with these gases is complicated since several peaks (the total energy peak plus associated escape peaks) result from a single x-ray energy. In addition, continuous purification during operation with hot metal purifiers is generally needed with the heavier rare gases. Most sources of krypton contain ^{85}Kr and this limits its use in low-level counting. The preparation and operation of gas proportional counters are discussed by Curran and Wilson (1965).

Proportional counters have been largely replaced by semiconductor detectors. They are, however, in use for measuring the x and gamma rays from plutonium and americium in living subjects (see, for example, Yaniv *et al.*, 1972).

6.2.3.3 *Semiconductor Detectors*. High-resolution lithium-drifted germanium Ge(Li) semiconductor detectors differentiate photopeaks of nearly equal energies, although even the largest detectors do not have high efficiencies because their present sensitive volumes are still limited. For example, a 65-cm^3 Ge(Li) detector has been reported with an absolute efficiency of 4.7 percent at 662 keV (Cooper, 1972), which may be compared to 28 percent for a 7.6 × 7.6-cm NaI (Tl) crystal, although the energy resolution was about 20 times better than the NaI(Tl) crystal. This greater resolution and a lower background results in an effective sensitivity comparable to that of the NaI(Tl) crystal. Furthermore, when the semiconductor detector is operated in an anticoincidence shield a greater Compton continuum reduction is obtained, which further improves the sensitivity for analyzing radionuclides in mixtures (Nielsen, 1972).

Phelps *et al.* (1968) have described an 18-cm^3 anticoincidence shielded Ge(Li) spectrometer for counting 200-ml samples that is capable of measuring, for example, 10 pCi ^{137}Cs, a value not possible with a 7.6 × 7.6-cm NaI(Tl) system. Many environmental samples contain much water and simple concentration techniques such as freeze drying are useful for reducing large sample volumes to the 200-ml size used in this system. By reducing 2 liters of cow's milk, a concentration of 5 pCi l^{-1} was measured.

The best way to determine the calibration parameters for measurement of a radionuclide in a given system is to measure a source of that radionuclide having a known disintegration rate, prepared from a standard. The calibration source must be prepared and measured in exactly the same way as the samples for which the calibration is being made. Standard solutions for many radionuclides may be obtained from the National Bureau of Standards. If a standard source is not available, calibration may be accomplished with a standard or standards of radionuclides emitting radiation similar in type and energy as a reference source. A long-lived reference source is convenient for calibration checks in place of short-lived nuclides which cannot be easily obtained or maintained. Although not normally a problem, periodic calibration is necessary to detect deterioration in counting efficiency.

The efficiency and energy calibration of a Ge(Li) detector is often made using only a few standard radionuclides; however, a large number covering the entire range are needed for measurements of energies to within 0.1 keV. The National Bureau of Standards now provides standards containing mixtures of radionuclides covering a wide energy range. Generally available gamma-ray standards suitable for spectrometer calibration are listed in Heath (1969). Calibration should be done with the same radionuclides to be measured or determined from an absolute efficiency curve. A convenient method by Äström *et al.* (1954) uses radionuclides which have two or more gamma-ray transitions with well known relative intensities. Temperature changes, and to some extent large variations in counting rates, can induce changes in amplification and change the channel in which events of a given energy are stored. This gain shift must be taken into account before applying calibration data or analyzing sample counting results. Least squares fitting of identified peaks in the spectrum is used to determine a new calibration from which corrections are derived for application to computer analysis of sample spectra. As an example of this technique, Pauly *et al.* (1966) report a library of 57 elements and the estimation of the detection limits of about 40 elements in graphite.

Computer analysis of gamma-ray spectral data is common and especially useful when many energy peaks are present. De Haan *et al.* (1965) discussed the applicability of such techniques to specific situations. An example of an extensive analysis system for NaI(Tl) and Ge(Li) spectrometry is given by Putnam *et al.* (1965) and more recently by Helmer and Putnam (1972). This program can analyze up to 300 peaks with as many as 50 calibration lines. A similar program for a small computer is given by Korthoven (1972). In these analyses, tests to detect large errors hidden in the computer code are desirable.

Ge(Li) detectors must always be kept at cryogenic temperatures to prevent precipitation of the drifted lithium. The promising high purity germanium detectors, however, do not require lithium compensation and, although they still must be operated at cryogenic temperatures, they may be safely stored at room temperature. Such detectors are commercially available in small sizes suitable for x-ray detection, and some larger sizes have been made and tested which are large enough for gamma-ray detection. For example, Llacer (1972) has developed a 32-cm^3 planar detector and two coaxial detectors of 2.4 cm^3 and 5.6 cm^3. The resolution of detectors made with high purity germanium is comparable to that of lithium drifted germanium detectors.

6.2.3.4 *X-Ray Detectors*. Israel *et al.* (1971) and Wogman (1975) have compared the energy resolutions and photopeak efficiencies of Si(Li) and Ge(Li) semiconductors, xenon proportional counters and NaI(Tl) detectors for measurement of x rays with energies between 10 and 300 keV. Figure 6-1 shows measured resolutions for versions of the four detectors. Resolutions for the Si(Li) and intrinsic Ge detectors were much better than 1 keV for 20-keV photons. Values for large Ge(Li) detectors, ranging from about 0.6 to 2 keV, are substantially better than for the older and smaller Ge(Li) detectors. Resolution of the xenon proportional counter, while very good for low energy photons, was only 10 keV for 100-keV photons.

The poorer resolution of the 7.6×7.6-cm NaI(Tl) detector, despite its greater counting efficiency, makes it unsuitable for spectrometry

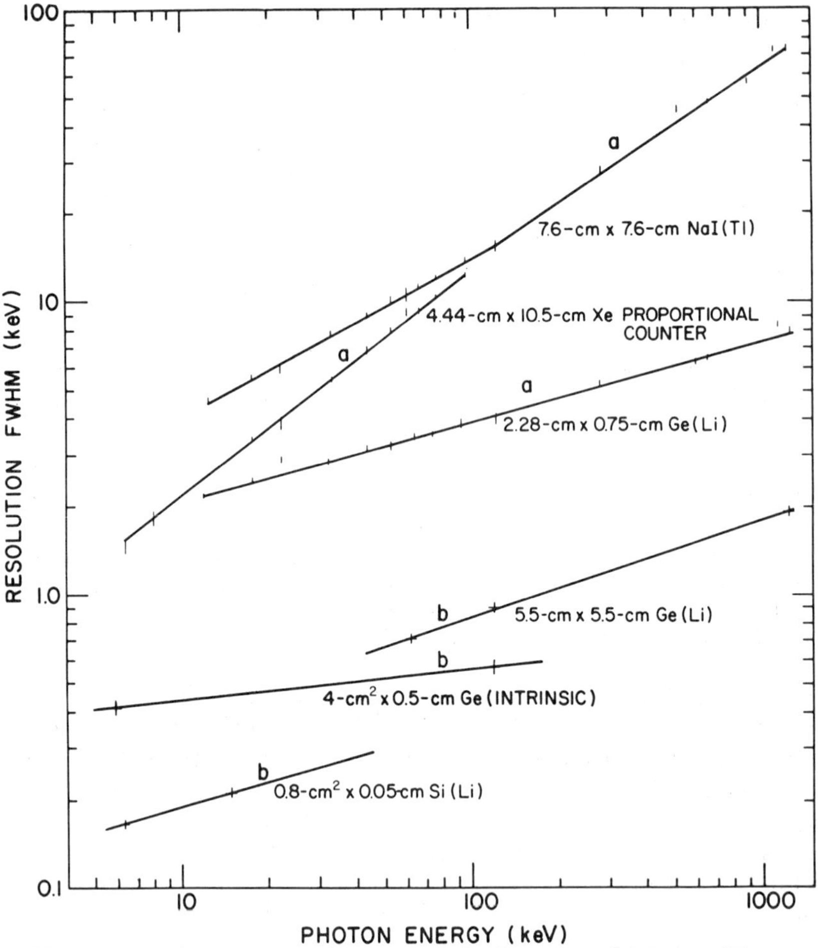

Fig. 6-1. Resolution versus photon energy for four types of detectors. Measurements were made with fluorescent x-rays and radionuclide gamma rays (a. from Israel, H. I., Lier, D. W. and Storm, E. (1971). "Comparison of detectors used in measurement of 10 to 300 keV x-ray spectra," Nucl. Instrum. Meth. **91**, 141, by permission; b. from Wogman, 1975).

below about 100 keV. Reported resolutions of thin NaI(Tl) detectors, such as those used for the *in situ* measurements of ^{241}Am described in Section 4.5.4, do not appear to be significantly better.

Figure 6-2, a comparison of calculated and measured inherent photopeak efficiencies of the four detector types, shows that the xenon proportional counter and silicon semiconductor have no role in spectrometry above 100 keV. The Si(Li) detector efficiency decreases rapidly to about 2 percent at 100 keV. The efficiency of the Ge(Li) detectors, a maximum at about 60–80 keV, steadily decreases with

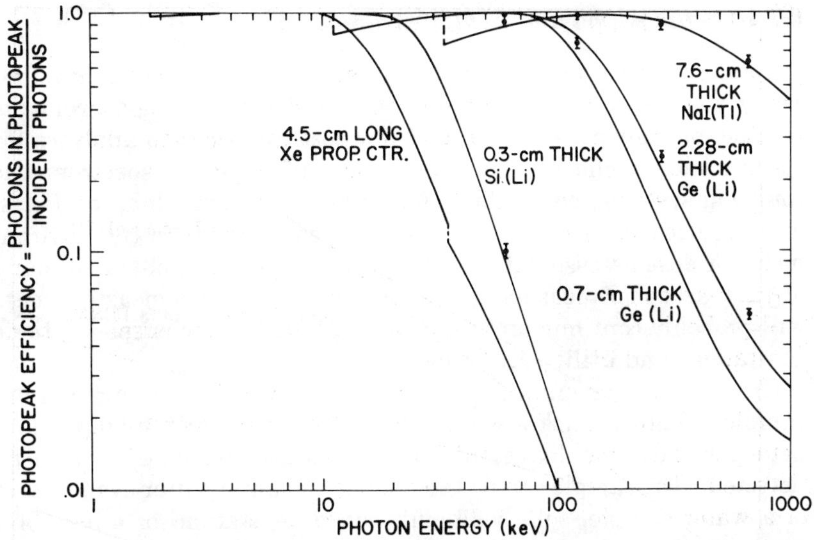

Fig. 6-2. Monte Carlo calculation of photopeak efficiency for the four detector types. Points indicate measured values [from Israel, H. I., Lier, D. W. and Storm, E. (1971). "Comparison of detectors used in measurement of 10 to 300 keV x-ray spectra," Nucl. Instrum. Meth. **91**, 141, by permission].

decreasing energy and above 100 keV drops rapidly. The large Ge(Li) detectors currently available have much higher efficiencies (about 25 percent for ^{137}Cs, compared to a 7.6 × 7.6-cm NaI(Tl) detector) than those determined by Israel *et al.* (1971). The 7.6 × 7.6-cm NaI(Tl) detector has the highest efficiency of any of the detectors, especially at the higher energies. At 300 keV, it is still 90 percent efficient, while a 2.28-cm thick Ge(Li) detector is only about 20 percent efficient. These data aid in selecting the best detectors for specific photon energy regions.

The excellent resolution and freedom from escape peaks make the Si(Li) detector the best for use below about 20 keV, although it is limited by low efficiency and relatively high Compton scatter at the higher energies. The Ge(Li) detector is best for the region between 30 and 100 keV, having excellent resolution, acceptable efficiency and relatively small escape peak losses, and it is still acceptable in the 100 to 3000 keV range. The NaI(Tl) detector is the best for the region between 100 and 300 keV if the spectra are not too complex. The advantage of the xenon proportional counter is that it can be made large to measure extended plutonium and americium sources in man or environmental samples. The ultra-pure, thin, large area configuration germanium semiconductor shows promise and is being investigated by Armantrout *et al.* (1974).

6.2.4 *Gross Alpha and Beta Measurements*

Total or "gross" activity measurements have been widely used for many years in making environmental radioactivity measurements because of their ease, speed, and low cost compared to analyses for specific radionuclides. The development of improved spectrometric counting systems, particularly for gamma rays and alpha particles, and the need for specific radionuclide measurement in the environment for dose assessment and pathway analysis, has obviated many of the reasons for measuring total activity. Since such measurements will probably continue and may have value for some purposes, their limitations and utility are discussed.

Gross measurements are commonly made by preparing a solid sample of known thickness and area and counting all the alpha and beta particles that leave this source with sufficient energy to be detected.[6] The sample can be the residue remaining after evaporation of a water sample, soil, sediment, ashed vegetation, or other solid material. Since the samples generally contain an unknown radionuclide mixture, an arbitrary choice must be made in selecting a standard for calibrating the counting system and determining the conversion of sample counting rates to disintegration rates. Calibration sources containing known amounts of the standard nuclide must be prepared that resemble the activity distribution, thickness, average atomic number, density and area of the samples to be counted. For samples of various thicknesses a counting efficiency curve as a function of weight or thickness should be determined.

The arbitrary but necessary choice of a standard nuclide introduces the largest uncertainty in the disintegration rates inferred from counting rates of thick samples, since correction factors such as self-absorption are energy dependent. The counting rate from an alpha-emitting source is proportional to the alpha particle range, which in turn varies as the 1.5 power of the energy. Since most alpha energies are between 4 and 8 MeV, the disintegration rate inferred from calibration with a 5 MeV alpha standard will usually agree with the true rate within a factor of two.

The end point energies of environmental betas vary from about 0.02 to 3 MeV. This corresponds to a difference in beta-ray range of about 1.5 g cm^{-2}, so the true disintegration rate inferred may be seriously in error. The meaning of the results obtained from gross activity measurements must therefore be clearly described; for example, the

[6] Equipment for gamma-ray spectrometry is currently so widely used and readily available that gross gamma-ray counting will not be discussed and is not recommended.

calculated disintegration rate refers only to the concentration of the standard nuclide that would result in the observed counting rate. No better claim can be made. That no information on the identity of the radionuclides is obtained is the most serious weakness of gross activity measurements. The lack of this information makes impossible any meaningful conclusions on the dosimetric significance of gross measurements, and on the environmental behavior and movement of the radionuclides in the sample.

The most serious abuse of such measurements is its use for dosimetric purposes or for determining compliance to some accepted or prescribed concentration.

The closer the resemblance between the radiations from the standard and from the unknown nuclides and the thinner the sample the more accurate the result. There is a practical maximum to the sample thickness, since the counting rate from a sample will remain constant once the thickness reaches the range of the radiations. The range of a 5-MeV alpha particle in a light element matrix such as soil is about 5 mg cm^{-2}. Since gross counting is commonly performed on a 5-cm diameter sample, the maximum weight is only about 100 mg; however, this is a quantity of material that is difficult to spread uniformly over 20 cm^2. Larger samples can be used if steps are taken to insure a homogeneous distribution of the alpha activity. Because of the large variation in beta particle energies, one should not generalize on maximum sample thickness.

The standard nuclide chosen should also be similar in energy to the expected radionuclides. Natural uranium would be the logical choice for samples expected to contain principally those alpha emitters, ^{239}Pu to interpret the results in terms of a specific alpha energy, or ^{137}Cs because it represents the average beta energy of an aged fission product mixture.

As an example of the accuracy possible, a standard procedure for water residues (APHA, 1971) was subjected to an interlaboratory comparison (Baratta and Knowles, 1971). The procedure requires counting of up to 200 mg of solids and a conductance measurement on the water sample to aid in selecting the sample size. Any finely divided solid material, e.g., soil or plant ash, can also be measured in this way. In the comparison, natural uranium and ^{137}Cs were used as the alpha and beta standards, respectively, and were also added as the unknown activities in the intercomparison samples. The results from about 15 laboratories averaged 10–15 percent low for alpha activity and were within 1–2 percent of the correct result for beta activity; a few individual laboratories were sometimes in error by as much as a factor of two. This study tested only the reproducibility and accuracy with which the activities of solid samples could be counted if the nuclides present were known.

The errors from gross activity measurements may increase when the standards and samples contain nuclides of different energies. In an unpublished study in which ^{239}Pu and ^{204}Tl were used as the standards for alpha and beta activity, respectively (Sedlet and Stehney, 1974), and natural uranium was the principal alpha emitter in water samples, the alpha disintegration rate inferred averaged about 85 percent of the true rate, similar to the results of the Baratta and Knowles (1971) study. When fission product mixtures were the principal beta emitters, the disintegration rates, inferred from a ^{204}Tl beta calibration, averaged up to 75 percent higher than the actual rate from a 100-mg cm^{-2} thick sample. The use of beta standards such as ^{137}Cs and ^{204}Tl will lead to large underestimates of the concentrations of very low energy beta emitters such as ^{14}C, ^{35}S, and ^{63}Ni. Separate analyses should be performed for such nuclides if they could be present and gross beta counting is used only as an approximate measure of total beta activity (Kahn *et al.*, 1971).

It should also be noted that if no information on the energy of the radiations is obtained, some of the alpha or beta radiations may be recorded as the other particles. Some energy information can be obtained from a beta absorption curve and, for alpha particles, from thick-sample alpha spectrometry (see Section 6.2.1), but the results should not be considered quantitative.

Gross activity counting should be reserved for surveying general radioactivity levels (if it is recognized which nuclides will be seriously over- or under-estimated) and for repeated rapid measurements of samples whose relative radionuclide composition is known and constant. It can also be used for verifying that the alpha and beta activities in a sample have been accounted for by specific radionuclide analyses. Gross counting results sometimes may not be interpretable. For example, no information on the plutonium concentration in air is obtained by total alpha counting of the long-lived activity in filter samples when the plutonium content is only about 1 percent of the total alpha concentration (Table 2-9).

Any improvement in gross alpha and beta measurements will probably result from a procedure that will produce particle energy information by counting with energy dependent detectors. Liquid scintillation counting of water samples has been used for this purpose by Bogen and Welford (1971). A single measurement of this type will give the low energy beta activity (^{3}H, ^{14}C, etc.), the medium to high energy beta activity (many fission and activation products), and the alpha activity. This technique is limited to very small sample sizes, which together with the relatively high background of liquid scintillation counters, limits the sensitivity. However, the technique may be useful for analyzing samples of nuclear facility effluents.

Many of the reasons formerly given for performing gross measurements, principally the savings in time and expense over radiochemical separations and analysis, are no longer applicable. Gamma-ray spectrometry has eliminated the need for radiochemical analysis of many nuclides, and when combined with some group chemical separations for very complex mixtures, will handle conveniently nearly all combinations of gamma emitters, so that fewer beta emitters need be analyzed separately. Improved separation methods and spectrometric counting techniques have simplified the analysis of alpha emitters substantially. The time saved by gamma-ray spectrometry should be spent on radiochemical analysis for those nuclides that cannot be determined by these techniques.

6.2.5 *Measurement of Radioactive Gases*

Radioactive gases can be measured in many types of counters and spectrometers, provided a suitable sample chamber is employed or the gas is placed inside the detector. Some techniques take special advantage of the gaseous nature of the samples for their measurement. These will be described for nuclides frequently measured in environmental samples, including radon, ^{85}Kr, ^{133}Xe, ^{14}C and ^{3}H.

Gaseous samples of beta or x-ray emitting radionuclides are commonly measured by use of internal gas counters or liquid scintillation counters in which the sample is introduced directly into the sensitive volume. Because of low cost and ease of automatic operation, liquid scintillation counting is usually chosen, although the more sensitive internal gas counters are needed for low-energy beta emitters. Of the three general types of internal gas counters, G.M. counters require the simplest electronics but do not have the capability of energy discrimination and have relatively long dead times. The relatively simple electronics for either current measurement or pulse counting with ionization chambers are offset by the higher backgrounds and lower sensitivities compared to proportional counters. The proportional counters are the most versatile of the internal gas counting systems; they are most suitable for low-energy beta counting and have low backgrounds and high sensitivities, with the best systems using anticoincidence shielding, massive shielding and pulse discrimination on both amplitude and rise time. Useful reviews have been made by O'Kelley (1968), Oeschger (1963), who described low level counting methods, and Schell (1971), who described methods for counting of tritium.

Internal gas counters are generally long metal, quartz or plastic cylinders having sensitive volumes of up to about 2 l. Sensitivity is further increased when they are built to be operated at pressures as high as 10 atmospheres. These counters, most frequently chosen when high sensitivity is required, are usually operated with an anticoincidence guard counter to distinguish the contribution of charged cosmic-ray particles to the background. Drever *et al.* (1957), and Houtermans and Oeschger (1958) used concentric cylinders for the counter-anticoincidence system and obtained a background of 0.6 min^{-1} for a 1.5-l detector. Recent interest in tritium and radioactive noble gases in the atmosphere has increased interest in even larger, higher pressure counters having rise time discrimination circuitry to enhance proportional counter sensitivities (Bradley and Willes, 1972).

Internal gas counters are suitable for all beta energies when the sample introduced is compatible with the counting gas. Radioactive noble gases are ideally suited to internal gas counting; this method is the most sensitive available, being able, for example, to determine ^{85}Kr at the present ambient level of about 10 pCi m^{-3} air (Sax *et al.*, 1968; Jaquish and Moghissi, 1973; and Johns, 1973). The detection limit for tritium (counted as HT gas) is about 10 pCi l^{-1} of water and for HT gas in atmospheric samples is about 2 pCi m^{-3} air (Östlund *et al.*, 1972). Enrichment may be required for determining the HTO in atmospheric rain and surface water samples.

The current status of liquid scintillation counting has been recently reviewed by Bransome (1970). In order to use this technique for counting radioactive gases, they must be soluble in the liquid scintillation medium. The noble gases have a high solubility in the toluene-based solvents, for example, 1 ml krypton being soluble in 1 ml solvent (Shuping *et al.*, 1969). In one method 10 ml krypton gas was dissolved in 25 ml of toluene-based liquid scintillator and ^{85}Kr concentration in air of about 10 pCi m^{-3} was measured to within about 4 percent accuracy. Tritium is usually converted to water and ^{14}C gas samples into benzene or a similar liquid for counting and are not generally counted as gases in liquid scintillators (Moghissi *et al.*, 1971).

In order to prepare a gas sample for counting it should be separated from the air and purified. This process can be very complex, consisting of separations by charcoal and molecular sieve columns and purification by drying and chemical reactions to remove impurities (Cummings *et al.*, 1971; Stevenson and Johns, 1971; and Johns, 1973).

A useful ^{85}Kr counting technique using solid plastic scintillators was developed by Sax *et al.* (1968). The method of placing ^{85}Kr in a 4-ml glass vial at a pressure of about 600 mm Hg[7] with plastic scintillator shavings is capable of a sensitivity of about 1 pCi m^{-3} air.

Standards for the calibration of these methods and instruments, available from the National Bureau of Standards, include ^{85}Kr gas, ^{3}H in water, and ^{14}C as oxalic acid.

Radon concentrations are determined by collecting the gas itself and counting it and its decay products, or by collecting and counting the decay products only. The decay products attach to particulate matter in air and are collected by filtration or, since they are principally positively charged, are collected on a negatively charged electrode. The simpler but less accurate technique is to collect and count the decay products rather than the gas itself. This question arises because radon recently emanated from the ground or solid objects is not in equilibrium with its decay products. Numerous articles on the measurement of radon by both techniques include a useful review on ^{222}Rn by Budnitz (1973). An older review by Sedlet (1966), while not specifically directed toward environmental monitoring, also considered the measurement of ^{220}Rn and ^{219}Rn.

To be measured at natural levels, ^{222}Rn must be concentrated from several liters of air passed through a cold (<0°C) charcoal collector. The preferred measurement method is alpha scintillation counting of the gas in the ZnS(Ag) phosphor coated chamber described by Lucas (1957) because it has been widely tested and used.

When the radon decay products are filtered and counted, the growth and decay equations for the radioactive chains must be correctly applied to the counting results (Thomas, 1968; 1972; and Martz *et al.*, 1969).

One measurement technique not widely reported, liquid scintillation counting of radon and its decay products, has certain advantages. These include the very high counting efficiency for alpha particles and high energy beta particles and the relatively high solubility of radon in the organic solvents used in liquid scintillation counting (Horrocks and Studier, 1964). Although the energy resolution of liquid scintillators is considerably poorer than that of solid detectors, a limited amount of spectrometry can be performed (Horrocks, 1973; Rolle, 1970). The latter investigator immersed a filter paper containing the radon decay products in a thin layer of liquid scintillator to discriminate better between alpha and beta particles.

The high counting efficiency for both alpha and beta particles

[7] 1 mm Hg ≈ 133 Pa (Pascal).

obtained by immersing filter paper containing radon decay products in a liquid scintillation medium was used by Assaf and Gat (1967) to measure the ^{214}Bi-^{214}Po pair (in the ^{222}Rn chain) and the ^{212}Bi-^{212}Po pair (in the ^{220}Rn chain) in the study of radon daughter disequilibrium in air.

The recent developments in noble gas chemistry may provide simple methods for removing radon and xenon from the air by oxidation to solid fluoride compounds that can be measured by relatively simple methods. A liquid fluorinating agent, bromine trifluoride, was used to remove radon effectively from air streams (Stein, 1973a) and a solid fluorinating reagent, dioxygenylhexafluoroantimonate (O_2SbF_6), was used in the laboratory to remove both ^{222}Rn and ^{133}Xe from an air stream with efficiencies approaching 100 percent at 23–25°C (Stein, 1973b). One of the difficulties with this technique is that the fluorinating agents as well as the noble gas fluorides react rapidly with water, so that the air must be thoroughly dried before the separation, and the collected xenon and radon kept dry.

6.2.6 *Mass Spectrometry*

Alpha counting procedures are capable of sufficient sensitivity to determine levels for monitoring and dose assessment. However, the potential of long-term buildup of these long-lived nuclides indicates that more sensitive and selective methods are required to determine the routes and rates of movement through the environment. Isotopic analysis is desired to identify the material, its source, and possible variables due to different and perhaps changing chemical and physical forms upon release and subsequent movement. For these purposes mass spectrometry provides a unique and valuable tool. Of specific interest in plutonium analysis are the related radionuclides, such as ^{240}Pu, since they emit alphas with energies that cannot be easily distinguished and hence one must often resort to mass spectrometry.

For environmental samples, mass spectrometers utilizing thermal ionization sources, multiple state (tandem) analyzers and ion counting detectors are generally used. Thermal ionization is chosen for its high specific ionization efficiency for the actinide elements and for its compatibility with the preferred solid sample forms. The relative levels of spectral interferences from organic and other impurities as well as ions from unwanted sources (i.e., so-called memory effects) can be minimized with this type of ion source. A typical filament is a "carbonized" rhenium V-type (McHugh, 1969), onto which a solid or solution sample is loaded. Hoffman *et al.* (1971), and others have noticed that chemically pure tracers have a higher ion efficiency, often by as much as an order of magnitude, over samples that have undergone extensive processing. Sample impurities introduced or not removed during sample processing can degrade the ionization efficiency and contribute to background. These effects are not well understood and represent an ultimate limitation of mass spectrometry. Analyses are made by the isotopic dilution method wherein a specific added isotope (spike) of the element undergoes the same separation and source preparation steps, so concentration measurements are independent of chemical yield and overall ionization efficiency.

The best mass spectrometers are capable of measuring about 10^7 atoms or about 4 fg of an actinide element. The corresponding ^{239}Pu decay rate is about 0.5×10^{-3} min^{-1}. The required sample size depends on the concentrations of trace isotopes. The ability of the spectrometer to distinguish the trace isotopes from the major ones is indicated by its abundance sensitivity. This is the ratio of peak intensity to the intensity of the tails, one mass unit on either side of the main peak. The tail contribution is the major background over which the other isotopes must be measured. Abundance sensitivity increases with the number of analyzer stages.

A single-stage machine operating in the actinide region at about 10^{-8} mm Hg has an abundance sensitivity of about 10^4 (Fields *et al.*, 1970), so that the tail contribution to the peak would be 0.01 percent. For two-stage (White and Collins, 1954; White and Forman, 1967) and three-stage (White *et al.*, 1958) machines, the abundance sensitivities would be about 2×10^6 and 5×10^8, respectively. Thus, in order to measure trace constituents to 1 ppm one would need a two or three stage mass spectrometer. A good measurement of the plutonium isotopes in environmental materials from world fallout can be made on samples as small as about 1 pg, corresponding to 0.1 min^{-1} plutonium. In addition to the ^{239}Pu, this plutonium has a composition of about 14 atom percent ^{240}Pu, 0.6 atom percent ^{241}Pu, and 0.3 atom percent ^{242}Pu.

Somewhat better performance is observed for uranium determinations because the filament is operated at several hundred degrees higher temperature to give a higher ion yield and a lower impurity interference than for plutonium.

Some typical applications of mass spectrometry to the measurement of actinide elements in environmental samples are: (a) plutonium in soil around a nuclear weapons facility (Krey and Hardy, 1970); (b) plutonium, americium, and curium in thermonuclear bomb debris (Diamond *et al.*, 1960); and (c) neptunium, uranium, and other actinides on the moon (Fields *et al.*, 1972). Fields *et al.* (1971) reported that lunar sample measurements made with a single stage 250-cm mass spectrometer built especially for the actinide region were able to establish an upper limit of 60 ag ^{244}Pu per gram lunar soil in a 3-g sample, with a sensitivity of about 200 ag for plutonium.

7. Conclusions: Trends and Problem Areas

7.1 Correlation of Research and Surveillance Data

If monitoring programs for nuclear facilities are improved, the data can serve as a large information source for both assessing dose from, and determining long-term trends in, environmental radioactivity. Facility operators should not carry the complete burden of performing the extensive low-level measurements and evaluations required for any general program, but such an effort can contribute to a fund of knowledge derived from government agencies' monitoring activities and scientists performing research.

The compilation and evaluation of monitoring results accumulated by the various organizations for different immediate purposes have not yet been accomplished. Such a meaningful compilation might be difficult to achieve; however, it would at least be prudent to assure that the sizeable effort for facilities monitoring does not continue to produce information of only short-term value and be relegated to the limited documentation required by regulations and standard practices. A successful effort in the accumulation of generally useful data from routine monitoring would require some planning, cooperation, and agreement between the participating institutions and organizations on the specific details of sampling and measurements. This report is an attempt to give some guidance in this direction.

Ground-level measurements made within a few miles of a particular nuclear plant by different organizations could be made complementary, in view of the expected increased use of nuclear power. These measurements could supplement airborne surveys made for monitoring larger areas or developing an emergency response program. Airborne measurements have already produced useful data on background terrestrial radiation and their continuance would provide indications of long-term trends in manmade radionuclide concentrations if carried out in a systematic way.

To respond properly to the concern about manmade radionuclides in the environment from a large nuclear industry requires data from a variety of monitoring techniques, of which airborne surveys are one of the most important. Relatively small changes, such as those that could produce 5 mrem per year to individuals in the general population, in the average terrestrial exposure rate over large areas from deposited radionuclides, can be identified by periodic surveys. The quantification of certain radionuclides is possible, but more effort should be directed toward determining individual nuclide concentrations for specific areas. The accumulation of these data, if done systematically and properly interpreted, can aid man in understanding and coping with environmental changes and allow him to locate a wider variety of natural resources for his benefit.

A further practical value of surveys can be realized in evaluating whether a flood potential exists due to the snowpack and determining large scale water equivalent information on a periodic basis (see Section 4.6.5). Observations, relatively limited at present, of anomalous U/Th ratios would be helpful for identifying and correlating important and valuable geologic environments with their *in situ* gamma-ray spectra. Improvements in measurement accuracy are required for soil moisture determinations and mineral exploration, so that instrument development and testing, and better interpretative techniques, are desirable.

Gamma-ray spectrometry requires more sensitivity and higher spectral resolution for airborne surveys than for ground-level measurements. The arrays of large volume scintillation detectors employed at present should be supplemented with high-energy resolution spectrometers. Increased airborne surveys with fixed-wing aircraft and helicopters in conjunction with ground measurements are desirable. Anomalies found in the aerial data can be verified by selected ground measurements.

7.2 Demands on Instrumentation and Interpretative Techniques

Much of the past interest in measurement programs in the United States and around the world was concerned with the development and maintenance of monitoring capabilities and computational methods for evaluating worldwide fallout. As a consequence, the state of sample collection and laboratory analyses is sufficiently reliable and sensitive for most determinations of radionuclide concentrations when made by experienced investigators, associated with well-equipped laboratories.

The experience in facilities monitoring has been based largely on performing measurements for evaluating the extent of unexpected radionuclide releases. The recent and additional interest in determining *de facto* radionuclide levels during normal operation has resulted in improvements in measurement reliability and sensitivity, due in part to the adoption of fallout measurement methods. However, the extremely low release rates and consequent low doses to members of the public permitted by regulatory organizations from relatively local sources emphasize the need for complementary measurements and calculations.

The expected wide use of plutonium in nuclear power production will place a large burden on environmental monitoring, which is based now almost completely on sample collection and laboratory analyses. It is prudent, therefore, to begin the development of supplementary detection methods, including possibilities for *in situ* measurements, to aid in this monitoring.

The various special problems associated with evaluating the health effects of inhalation of particles containing alpha emitters, such as the transuranic elements, also indicate a requirement for instruments capable of measuring these materials. Not only should the amount of the radionuclide be determined, but in many cases, its chemical and physical form. Dose assessment requires information regarding the release of these materials to the environment, their characterization, subsequent changes during their transport through the environment, deposition, and resuspension. The determination of all of these factors requires measurements that are now made with difficulty and generally at great expense. Instruments for better, faster, more economical, and real-time operation would aid the present methods of sample collection and analysis.

Spectrometric methods patterned after those described in Section 4.5.3 for fallout and natural emitters, if successful, can aid in making determinations of any local deposits, or their absence. Qualitative survey methods similar to those employed with the FIDLER, as described in Section 4.5.4, can aid in determining optimum environmental sampling locations and consequently reducing sample collections.

Although *in situ* alpha counting measurements are generally not practical, one possibility involves alpha counting of carefully collected soil samples arranged in a fixed geometry plate. The expected alpha activity from natural radionuclides can be estimated from *in situ,* high resolution spectrometer determinations of the ^{238}U and ^{232}Th series concentrations and the excess alpha activity thereby attributed to manmade additions. Another technique could rely on the placement of large numbers of track etch plastic films to record alpha tracks, although interference by radon daughter alpha tracks is a problem.

The already available large coaxial Ge(Li) detectors provide pulse height spectra with a low energy limit of about 50 keV. These spectra exhibit gamma-ray peaks associated directly with ^{239}Pu, in addition to the prominent 60-keV peak due to ^{241}Am. Studies are needed to determine the practical usefulness of this spectrometry for quantifying ^{239}Pu concentrations. These studies should include correlative determinations of transuranic nuclide compositions and distributions in suitable environmental samples and subsequent data interpretation.

Field studies with arrays of intrinsic germanium or scintillation detectors are needed to evaluate advanced FIDLER concepts with detailed analyses of *in situ* spectra from a few to about 100 keV. These high resolution spectra may provide the guidance needed to separate ^{239}Pu gamma rays and de-excitation x rays from similar natural and manmade radiation.

In a similar fashion, field monitoring of ^{85}Kr released from the nuclear fuel cycle is needed. Direct, continuous measurements of dose rate or ground-level concentrations near inhabited areas would supplement tedious sample collection methods.

Monitoring of other relatively long-lived radionuclides, such as ^{129}I from reactor fuel reprocessing, will probably continue to be based on periodic sample collection and analysis. To identify any trends in the concentrations of such nuclides will require development of sensitive, reliable, and inexpensive sampling and analytical methods, as well as high quality analysis.

Continuous gamma-ray monitors, either integrating detectors such as thermoluminescence dosimeters or instruments with responses which are approximately independent of gamma-ray energy, are now capable of documenting environmental gamma-ray variations and identifying those due to plant operation. Some of these instruments are based on past extensive work by cosmic-ray physicists, geophysicists, and geochemists. More powerful monitors employing spectrometers can probably now be built and operated, although their wide use in facilities monitoring is probably unjustified, their periodic use should be helpful and may provide guidance on the location and types of monitors to be employed, as well as on modifications that need to be applied to monitoring programs.

7.3 Development of Nuclear Facilities Surveillance Programs

Past guidance in the design and conduct of nuclear facilities monitoring programs was general, although efforts by regulatory organizations in the 1970's seem to have encouraged facilities operators to improve monitoring. While these efforts, and previous efforts by professional societies, have led to the promulgation of selected standard analytical methods, systematic guidance is still limited to reports published by the ICRP and the EPA (ICRP, 1965a; EPA, 1972). The type of guidance to be provided for the nuclear industry should be based on needs identified by interested parties (regulatory organizations, plant operators, and expert advisory groups or standards-making bodies) and must be tempered by considerations of the exact purposes of monitoring and by the cost of monitoring programs.

The type of facility to be monitored will determine the measurement methods required, and measurements of limited or zero value should be avoided. The lowest practical values of dose or radionuclide concentrations reflected by recent pronouncements of regulatory organizations, for example, indicate that measurements of so-called gross radioactivity in soil, water, and air, made in the presence of natural radioactivity, should not be considered useful for determining dose or compliance with allowed releases. On the other hand, the cost of desired monitoring makes evident the practical need for making measurements on a selective basis.

Three general bases for monitoring, neglecting relatively large radionuclide releases, can be identified. Measurements are made to aid in dose assessment, for the determination of compliance with allowed releases, or for the identification of long-term trends of environmental concentrations of particular radionuclides. The latter type of monitoring does not fall into the area of responsibility of nuclear facilities operators, although their data may represent useful contributions to the more general monitoring of trends that government agencies or researchers should do. Monitoring general radionuclide trends is discussed in Section 7.4.

Measurements made for dose assessment purposes necessarily supplement computations and some should provide benchmark data against which computations can be tested. Computational models should be modular in character, so that testing of at least some of the components is possible, and reliance on untested models should be avoided. Ideally, reliance on tested models can result in the reduction of measurements in an environmental monitoring program, but limited measurements selected for the particular facility should be continued to aid in interpreting changes which may arise from plant and terrain modification.

As was indicated in Section 3, external radiation dose assessment measurements are simpler and more direct than those made for determining internal dose due to the movement of radionuclides to man, for example, through his food. Dose assessment of radionuclide pathways is complex and further investigations on how to simplify such complex dose assessments should be made. Such investigations should concern both the improvement of our knowledge on pathways and the identification of particular measurements required for routine monitoring.

Compliance with allowed releases usually depends on the plant operator's measurements of the quantities of particular radionuclides released in stipulated time intervals. Traditionally, such measurements are made at each of the few release points purposefully designed into the plant. In plants designed to hold up or delay releases, however, dispersion of radionuclides into the environment may result from general, low-level leakage and not from releases from known locations. To assure that any remaining releases are, in fact, inconsequential, supplementary environmental measurements may be required to at least provide desirable benchmark data.

7.4 Assessment of Long-term Trends

The assessment of long-term trends should produce information on the redistribution of radioactivity. Long-term measurements will be a necessary function in a nuclear-power society as a part of assessing changes in man's environment. The time and spatial variation in worldwide concentrations of certain long-lived nuclides, such as ^{85}Kr and ^{129}I, and possibly other volatile long-lived radionuclides, will need study, and localized concentrations of many long-lived radionuclides near fuel reprocessing plants and radioactive waste storage facilities will require long-term monitoring.

Monitoring will also be needed to assess any long-term or slow changes in the forms of nuclides (such as plutonium) in the environment that may affect their biological availability or their environmental pathways. Such factors as resuspension, solubilization, and complex formation in soil and in the beds of bodies of water will have to be assessed as a function of time. This assessment will be more difficult than monitoring for concentration alone and may require the development and use of specialized measurement techniques and data interpretation methods.

The three-fold conditions for a successful national or international effort for determining long-term trends are suitable measurements, i.e., those designed to detect the particular radionuclides of interest; the development and field testing of new measurement methods, some of which have been indicated in this report; and the periodic assessment of the data by an independent authoritative body. The key to a successful long-term assessment would be the organization of a small, permanent scientific secretariat which can infrequently convene a larger body of experts.

7.5 Redistribution of Natural and Manmade Radioactivity

One of the factors that affects human population exposure to natural background radiation is the redistribution of naturally-occurring radionuclides in the environment. As a consequence of modern civilization, redistribution takes place as a result of topsoil modification, water treatment in agriculture, housing construction, and industry.

The outdoor radiation level at a particular location is a time-varying function of various environmental parameters, such as the radionuclide content of the upper soil and lower atmosphere, water content of the ground, and density of foliage. However, over time periods sufficiently long that the effect of variations in the local topography are averaged out, the mean radiation level is characteristic of that location. If topsoil or fertilizers from another location are applied or if the ground is plowed, the radionuclide content of the upper soil layers may be significantly changed. The addition of water to the ground decreases the effective radionuclide concentration in the ground and attenuates the radiation from the ground. The presence of a structure attenuates the outdoor radiation while producing a radiation field of its own from any radionuclides contained in the construction materials. Industrial activities can result in the emission of natural radionuclides to the air that can be deposited on the ground.

All of these phenomena have been noted in various radiation measurement programs. For example, Fryer and Adams (1974) have suggested that radiation arising from excess radium in phosphate fertilizer is responsible for anomalous U/Th ratios observed in helicopter-borne spectrometric measurements over plowed fields. Jaworoski *et al.* (1972) have estimated that 400 Ci of ^{226}Ra are dispersed annually over plowed fields around the earth in the form of such fertilizer. The same authors have estimated that 150 Ci of the same nuclide are dispersed to the earth's atmosphere as a result of industrial coal combustion, and have documented decreasing ^{226}Ra concentrations in snow as a function of distance from a coal-burning facility.

The effect of housing on the exposure of human populations has been discussed by Oakley (1972) and NCRP (1975a). Typical effective attenuation factors for various types of construction materials, taking into account both the actual attenuation of the outdoor radiation field in materials and the field produced by radionuclides in these materials, have been estimated and applied to the assessment of population exposure in urban areas. However, these estimates have been based on extremely limited experimental data that may not be generally applicable. An estimate of population exposure based on consideration of the indoor radiation environment as a perturbation of the outdoor environment is not necessarily an acceptable approach unless indoor data are totally inadequate.

Any redistribution of long-lived manmade radionuclides may follow the same pattern as the natural. The worldwide deposition of fallout radionuclides increased the outdoor gamma-ray levels in a way which is somewhat analogous to the widespread use of radium- or potassium-rich fertilizers. In like fashion, the dispersion in the atmosphere of manmade radionuclide emissions from nuclear facilities is similar to the radium-rich flyash from coal-burning facilities.

From the standpoints of research and determining trends, we should consider the development of a radiation monitoring capability that not only provides information on man's present radiation exposure and its sources, but how the total exposure levels and the source properties are changing with time as a result of man's various activities. It is sometimes forgotten that exposure to *both* natural and manmade radiation sources is dependent on the nature of these activities, and thus may be at least partly controllable. The following types of investigations would contribute needed empirical information:

(a) studies of industrial emissions to the environment and their quantitative contribution to radiation exposure;
(b) studies of the effect of soil additives applied in populated areas on outdoor radiation exposure and radon levels;
(c) studies of radioactivity in widely-used construction materials and its relationship to the origin of the raw material:
(d) studies of indoor radiation fields, including radon levels, and their dependence on source concentration and spatial distribution; and
(e) studies of population exposure in selected areas, and its dependence on local source distributions, housing, and living habits, and possibly also on the consumption of food prepared at distant locations.

Many of the measurement techniques discussed in the preceding sections of this report will find immediate application in such studies. The measurement of both the instantaneous amplitude of environmental radiation levels or radionuclide concentrations, and their time variations, as well as long-term perturbations in these values, are within the capabilities of these techniques.

REFERENCES

ABERG, B. AND HUNGATE, F. P., Eds. (1967). *Radioecological Concentration Processes* (Pergamon Press, Oxford).

ADAMS, J. A. S. (1962). "Radioactivity of the lithosphere," page 1 in *Nuclear Radiation in Geophysics*, Israel, H. and Krebs, A., Eds. (Academic Press, New York).

ADAMS, J. A. S. (1969). "Total and spectrometric gamma-ray surveys from helicopters and vehicles," page 147 in *Nuclear Techniques and Mineral Resources* (International Atomic Energy Agency, Vienna).

ADAMS, J. A. S. AND FRYER, G. E. (1964). "Portable gamma-ray spectrometer for field determination of thorium, uranium and potassium," page 577 in *The Natural Radiation Environment*, Adams, J. A. S. and Lowder, W. M., Eds. (University of Chicago Press, Chicago).

ADAMS, J. A. S. AND LOWDER, W. M., Eds. (1964). *The Natural Radiation Environment* (University of Chicago Press, Chicago).

ADAMS, J. A. S. AND GASPARINI, P. (1970). *Gamma-Ray Spectrometry of Rocks* (Elsevier Publishing Co., Amsterdam).

ADAMS, J. A. S., OSMOND, J. K. AND ROGERS, J. J. W. (1959). "The geochemistry of uranium and thorium," Phys. and Chem. of the Earth **3**, 299.

AITKEN, M. J. (1968). "Low-level environmental radiation measurements using natural calcium fluoride," page 281 in *Proc. of Second Intl. Conf. on Luminescence Dosimetry*, Auxier, J. A., Becker, K. and Robinson, E. M., Eds., Report CONF-680920 (U. S. Atomic Energy Commission, Washington).

ALTSHULER, B. AND PASTERNACK, B. (1963). "Statistical measures of the lower limit of detection of a radioactivity counter," Health Physics **9**, 293.

ANDELMAN, J. B. AND ROZZELL, T. C. (1970). "Plutonium in the water environment. I. Characteristics of aqueous plutonium," page 18 in *Radionuclides in the Environment*, Gould, R. F., Ed. (American Chemical Society, Washington).

ANDERSON, H. R. (1961). *Primary Cosmic Radiation in 1958 and Variations*, Thesis (California Institute of Technology, Pasadena, California).

ANSPAUGH, L. R., PHELPS, P. L., GUDIKSEN, P. H., LINDEKEN, C. L. AND HUCKABAY, G. W. (1972). "The in situ measurement of radionuclides in the environment with a Ge(Li) spectrometer," page 279 in *The Natural Radiation Environment II*, Adams, J. A. S., Lowder, W. M. and Gesell, T., Eds., Report CONF-720805 (U. S. Energy Research and Development Administration, Washington).

APHA (1971). American Public Health Association, *Standard Methods for the Examination of Water and Waste Water*, 13th ed. (American Public Health Association, Washington).

ARMANTROUT, G. A., WICHNER, R. AND SWIERKOWSKI, S. (1974). "Encapsulation of high-purity germanium detectors," IEEE Trans. on Nucl. Sci. **NS-21,** 344.

ARMS (1973). *Aerial Radiological Measuring System—Systems and Procedures Employed Through FY-71,* Report L-1038, EGG-1183-1526 (E. G. & G., Inc., Las Vegas, Nevada).

Artna-Cohen, A. (1971). "Nuclear data sheets for A = 239," Nucl. Data Sheets **B6,** 577.

Assaf, G. and Gat, J. R. (1967). "Direct determination of short-lived radon daughter products on air filters by liquid scintillation counting using a delayed-coincidence technique," Nucl. Instrum. Meth. **49,** 29.

ASTM (1973). American Society for Testing and Materials, *1973 Annual Book Standards, Part 23, Water; Atmospheric Analysis* (American Society for Testing and Materials, Philadelphia).

Äström, B., Wapata, A. H., Thulin, S. and Bergstrom, I. (1954). "Calibration of a scintillation spectrometer for the comparison of γ- and x-ray intensities," Arkiv Fysik **7,** 247.

Aten, A. H. W., Jr., Heertje, I. and DeJong, W. M. C. (1961). "Measurements of low-level environmental radiation by means of Geiger-Müller counters with observations in the Amsterdam region," Physica **27,** 809.

Attix, F. H., Ed. (1967). *Luminescence Dosimetry,* Symposium Series 8 (U. S. Atomic Energy Commission, Washington).

Auxier, J. A., Becker, K. and Robinson, E. M., Eds. (1968). *Proc. of Second Intl. Conf. on Luminescence Dosimetry,* Report CONF-680920 (U. S. Atomic Energy Commission, Washington).

Bailey, R. A. (1962). *Paper Chromatographic and Electromigration Techniques in Radiochemistry,* Report NAS-NS-3106 (National Technical Information Service, U. S. Department of Commerce, Springfield, Virginia).

Baines, M. E. D. (1963). *Determination of Plutonium Alpha Activity in Urine, Faeces, and Biological Materials,* Report AEEW-R-292 (U. K. Atomic Energy Authority, England).

Baranov, V. I. and Morozova, N. G. (1973). "The behavior of natural radionuclides in soil," page 3 in *Radioecology,* Klechkovskii, V. M., Polikarpov, G. G. and Aleksakhin, R. M., Eds. (Halsted Press, New York).

Barasch, G. E. and Beers, R. H. (1971). *Aerial Radiological Measuring Surveys of the Nuclear Fuel Services Plant, West Valley, New York, 1968 and 1969,* Report ARMS-68.6.9 (E. G. & G., Inc., Las Vegas, Nevada).

Baratta, E. J. and Knowles, F. E., Jr. (1971). "Statistical limits of accuracy and precision of gross alpha and beta radioactivity measurements in water," Radiological Health Data and Reports **12,** 195.

Bärring, N. E. (1966). *Electrodeposition of Actinide and Lanthanide Elements,* Report AE-217 (Actiebologet Atomenergie, Stockholm, Sweden).

Beck, C. K. (1971). "Environmental monitoring for USA licensed power reactors," page 401 in *Proc. of Symp. on Environmental Aspects of Nuclear Power Stations* (International Atomic Energy Agency, Vienna).

Beck, H. L. (1966). "Environmental gamma radiation from deposited fission products, 1960–1964," Health Physics **12,** 313.

Beck, H. L. (1971). "Monte Carlo calculations of electromagnetic shower transition effects," Nucl. Instrum. Meth. **91,** 525.

BECK, H. L. (1972a). *Absolute Intensities of Gamma Rays from the Decay of ^{238}U and ^{232}Th,* Report HASL-262 (U. S. Atomic Energy Commission, New York).

BECK, H. L. (1972b). "The physics of environmental radiation fields," page 101 in *The Natural Radiation Environment II,* Adams, J. A. S., Lowder, W. M. and Gesell, T., Eds., Report CONF-720805 (U. S. Energy Research and Development Administration, Washington).

BECK, H. L. AND DE PLANQUE, G. (1968). *The Radiation Field in Air Due to Distributed Gamma-Ray Sources in the Ground,* Report HASL-195 (U. S. Atomic Energy Commission, New York).

BECK, H. L. AND FREESWICK, D. C. (1975). "A continuous monitoring system for ^{85}Kr in the environment," Trans. Am. Nucl. Soc. **21,** 92.

BECK, H. L., CONDON, W. J. AND LOWDER, W. M. (1964). *Spectrometric Techniques for Measuring Environmental Gamma Radiation,* Report HASL-150 (U. S. Atomic Energy Commission, New York).

BECK, H. L., LOWDER, W. M., BENNETT, B. G. AND CONDON, W. J. (1966). *Further Studies of External Environmental Radiation,* Report HASL-170 (U. S. Atomic Energy Commission, New York).

BECK, H. L., LOWDER, W. M. AND MCLAUGHLIN, J. E. (1971). "*In situ* external environmental gamma-ray measurements utilizing Ge(Li) and NaI(Tl) spectrometry and ionization chambers," page 499 in *Rapid Methods for Measuring Radioactivity in the Environment* (International Atomic Energy Agency, Vienna).

BECK, H. L., DE CAMPO, J. A., GOGOLAK, C. V., LOWDER, W. M., MCLAUGHLIN, J. E. AND RAFT, P. D. (1972a). "New perspectives on low level environmental monitoring around nuclear facilities," Nucl. Tech. **14,** 232.

BECK, H. L., DE CAMPO, J. A. AND GOGOLAK, C. V. (1972b). *In Situ Ge(Li) and NaI(Tl) Gamma-Ray Spectrometry,* Report HASL-258 (U. S. Atomic Energy Commission, New York).

BECKER, K. (1972a). "Applied dosimetry research," page 62 in *Health Physics Division Annual Report,* Report ORNL-4811 (Oak Ridge National Laboratory, Oak Ridge, Tennessee).

BECKER, K. (1972b). "Environmental monitoring with TLD," Nucl. Instrum. Meth. **104,** 405.

BECKER, K. (1973). *Solid State Dosimetry* (Chemical Rubber Co., Cleveland, Ohio).

BELOUSOVA, I. M. AND SHTUKKENBERG, YU. M. (1961). *Estestvennya Radioaktivnost'* [Natural Radioactivity] (Medgiz, Moscow).

BENNETT, B. G. (1970). "Estimation of gonadal absorbed dose due to environmental gamma radiation," Health Physics **19,** 757.

BENNETT, B. G. AND BECK, H. L. (1967). *Legendre, Tschebyscheff, and Half Range Legendre Polynomial Solutions of the Gamma Ray Transport Equation in Infinite Homogeneous and Two Media Plane Geometry,* Report HASL-185 (U. S. Atomic Energy Commission, New York).

BERNAS, B. (1968). "A new method for decomposition and comprehensive analyses of silicates by atomic absorption spectrometry," Anal. Chem. **40,** 1682.

BERNAS, B. (1973). "Acid pressure decomposition device for interference-free

AA analyses," American Laboratory 5, 41.

BINDER, W., DISTERHOFT, S. AND CAMERON, J. R. (1968). "Dosimetric properties of CaF_2Dy," page 43 in *Proc. of Second Intl. Conf. on Luminescence Dosimetry*, Auxier, J. A., Becker, K. and Robinson, E. M., Eds., Report CONF-680920 (U. S. Atomic Energy Commission, Washington).

BLACK, C. A., Ed. (1965). *Methods of Soil Analysis* (American Society of Agronomy, Madison, Wisconsin).

BLANCHARD, R. L., KRIEGER, H. L., KOLDE, H. E. AND KAHN, B. (1971). "Radiological surveillance studies at a BWR nuclear power station—estimated dose rates," page 372 in *Health Physics Aspects of Nuclear Facilities Siting*, Voilleque, P. C. and Baldwin, B. R., Eds. (Health Physics Society, Idaho Falls, Idaho).

BLAND, C. J. AND SCHOENAICH-PETERS, R. (1972). "Feasibility study of the use of lead loaded plastic scintillator in airborne mineral prospecting by gamma-ray spectrometry," page 663 in *The Natural Radiation Environment II*, Adams, J. A. S., Lowder, W. M. and Gesell, T., Eds., Report CONF-720805 (U. S. Energy Research and Development Administration, Washington).

BOGEN, D. C. AND WELFORD, G. A. (1971). "Application of liquid scintillation spectrometry for total beta and alpha assay," page 383 in *Rapid Methods for Measuring Radioactivity in the Environment* (International Atomic Energy Agency, Vienna).

BONDARENKO, G. M., BAKSHEEVA, I. P. AND DUBROVA, L. N. (1971). "Ion-exchange purification of ^{90}Y in the analyses of ^{90}Sr in samples of sea water," Radiokhimiya 13, 314 (English Translation by Consultants Bureau, New York).

BONI, A. L. (1963). "Rapid determination of radioiodine-131 in milk," Health Physics 9, 1035.

BONI, A. L. (1966). "Rapid ion-exchange analysis of radiocesium in milk, urine, sea water, and environmental samples," Anal. Chem. 38, 89.

BONNER, N. A. AND KAHN, M. (1951). "Behavior of carrier free tracers," page 102 in *Radioactivity Applied to Chemistry*, Wahl, A. C. and Bonner, N. A., Eds. (John Wiley and Sons, New York).

BOSSHART, R. E. AND YOUNG, R. K. (1972). "Preparation of gas samples for liquid scintillation counting of carbon-14," Anal. Chem. 44, 1117.

BOWMAN, W. W. AND MACMURDO, K. W. (1974). "Radioactive-decay gammas ordered by energy and nuclide," Atomic Data and Nuclear Data Tables 13, 89.

BRADLEY, W. E. AND PEARSON, J. E. (1970). "Aircraft measurements of the vertical distribution of radon in the lower atmosphere," J. Geophys. Res. 75, 5890.

BRADLEY, A. E. AND WILLES, E. H. (1972). *A Risetime Discriminator for Low-Level Tritium Counting*, Report UCRL-73978 (Lawrence Livermore Laboratory, Livermore, California).

BRANDT, W. (1960). *Energy Loss and Range of Charged Particles in Compounds*, Research Report, unpublished (Radiation Physics Laboratory, E. I. duPont de Nemours & Company, Inc.).

BRANSOME, E. D. (1970). *The Current Status of Liquid Scintillation Count-*

ing (Grune and Stratton, New York).

BRINCK, W., GROSS, K., GELLS, G. AND PARTRIDGE, J. (1975). "Special field study at the Vermont Yankee nuclear power station," Personal Communication (U. S. Environmental Protection Agency, Cincinnati, Ohio).

BRODA, E. AND SCHONFELD, T. (1966). *The Technical Applications of Radioactivity,* Vol. I (Pergamon Press, Oxford).

BUDNITZ, R. J. (1973). "Radon-222 and its daughters," *RAD-NUC Instrumentation for Environmental Monitoring* (Lawrence Berkeley Laboratory, Berkeley, California).

BURCH, P. R. J., DUGGLEBY, J. C., PLDROYD, B., AND SPIERS, F. W. (1964). "Studies of environmental radiation at a particular site with a static γ-ray monitor," page 767 in *The Natural Radiation Environment,* Adams, J. A. S. and Lowder, W. M., Eds. (University of Chicago Press, Chicago).

BUREAU OF THE CENSUS (1963). "Consumption of selected food items in U. S. households," Radiol. Health Data **4,** 124.

BURKE, G. DE P. (1972). *Investigations of CaF_2:Mn Thermoluminescent Dosimetry System for Environmental Monitoring,* Report HASL-252 (U. S. Atomic Energy Commission, New York).

BURKE, G. DE P. (1975). *Variations in Natural Environmental Gamma Radiation and Its Effect on the Interpretability of TLD Measurements Made Near Nuclear Facilities,* Report HASL-289 (U. S. Energy Research and Development Administration, New York).

BURKE, G. DE P. AND MARCIN, D. G. (1974). *Computer Program for Calculating Various Parameters Related to Soil-Water Balance,* Report HASL-282 (U. S. Atomic Energy Commission, New York).

BURKE, G. DE P. AND MCLAUGHLIN, J. E. (1974). "Performance criteria for environmental radiation monitoring with TLD systems," IEEE Trans. on Nucl. Sci. **NS-21,** 444.

BURLIN, T. E. (1966). "A general theory of cavity ionization," Brit. J. Radiol. **39,** 727.

BURLIN, T. E. (1968). "Cavity-chamber theory," page 331 in *Radiation Dosimetry*, Vol. 1, Attix, F. H. and Roesch, W. C., Eds. (Academic Press, New York).

BURLIN, T. E. AND CHAN, F. K. (1967). "Some applications of cavity theory to condensed-state radiation dosimetry," page 393 in *Solid State and Chemical Radiation Dosimetry in Medicine and Biology* (International Atomic Energy Agency, Vienna).

BURSON, Z. G. AND FRITZSCHE, A. E. (1972). *Snow Gaging by Airborne Radiological Surveys—Status Through September 1972,* Report L-1078, EGG-1183-1565 (E. G. & G., Inc., Las Vegas, Nevada).

BURSON, Z. G., BOYNS, P. K. AND FRITZSCHE, A. E. (1972). "Technical procedures for characterizing the terrestrial gamma radiation environment by aerial surveys," page 559 in *The Natural Radiation Environment II,* Adams, J. A. S., Lowder, W. M. and Gesell, T., Eds., Report CONF-720805 (U. S. Energy Research and Development Administration, Washington).

BUTLER, F. E., LIEBERMAN, R., STRONG, A. B. AND MOSS, V. R. (1971). "Sampling and analysis of soils for plutonium," page 47 in *Proc. of an Environmental Plutonium Symposium*, Report LA-4756, Fowler, E. B.,

Henderson, R. W. and Milligan, M. F., Eds. (Los Alamos Scientific Laboratory, Los Alamos, New Mexico).

Cameron, J. R., Suntharalingam, N. and Kenney, G. N. (1968). *Thermoluminescent Dosimetry* (University of Wisconsin Press, Madison, Wisconsin).

Carmichael, H. (1971). "Comments on the absolute ionization of cosmic radiation at balloon altitudes covering four solar maxima," page 779 in *Proc. of Twelfth Intl. Cosmic Ray Conference* (University of Tasmania, Hobart).

Carmichael, H. and Bercovitch, M. (1969). "Analysis of IQSY cosmic-ray survey measurements," Can. J. Physics **47,** 2073.

Carmichael, H., Bercovitch, M. and Steljes, J. F. (1967). "Introduction of meteorological corrections into meson monitor data," Tellus **19,** 143.

Cassidy, M. E., Watnick, S., Negro, V. C., Freeswick, D. C. and Graveson, R. T. (1974). "A computer compatible field monitoring system," IEEE Trans. on Nucl. Sci. **NS-21,** 461.

Chesslet, M., Nordeman, D., Lalou, C. and Perrin, F. (1965). "Étude comparative dex traux de retombées radioactives maritimes et continentales," Comp. Rend. Acad. Sci. **260,** 2875.

Chester, J. P., Chase, R. L. and Wood, S. (1972). *A Digital Environmental Monitor,* Report BNL-16922 (Brookhaven National Laboratory, Upton, L. I., New York).

Choppin, G. R. (1961). *Experimental Nuclear Chemistry* (Prentice-Hall, Englewood Cliffs, New Jersey).

Chu, N. Y. (1971). "Plutonium determination in soil by leaching and ion-exchange separation," Anal. Chem. **43,** 449.

Clay, J. (1939). "The absolute value of cosmic-ray ionization at sea level in various gases," Rev. Mod. Physics **11,** 123.

Cline, J. E. and Heath, R. L. (1967). *Annual Progress Report, Gamma-Ray Spectrometry of Neutron-Deficient Isotopes,* Report IDO-17222 (U. S. Atomic Energy Commission, Idaho Falls, Idaho).

Cochran, W. G. (1963). *Sampling Techniques* (John Wiley and Sons, New York).

Cooper, J. A. (1972). "A versatile Ge(Li)-NaI(Tl) coincidence-anticoincidence gamma-ray spectrometer for environmental and biological problems," Nucl. Instrum. Meth. **99,** 125.

Coryell, C. D. and Sugarman, N., Eds. (1951). *Radiochemical Studies: The Fission Products* (McGraw-Hill, New York).

Cosolito, F. J., Cohen, N. and Petrow, H. G. (1968). "Simultaneous determination of iron-55 and stable iron by liquid scintillation counting," Anal. Chem. **40,** 213.

Cox, W. M., Blanchard, R. L. and Kahn, B. (1970). "Relation of radon concentration in the atmosphere to total moisture detention in soil and atmospheric thermal stability," page 436 in *Advances in Chemistry Series No. 93* (American Chemical Society, Washington).

Crouthamel, C. E. and Heinrich, R. R. (1971). "Radiochemical separations," Part I, page 5467 in *Treatise on Analytical Chemistry,* Kolthoff, I. M. and Elving, P. J., Eds. (John Wiley & Sons, New York).

CUMMINGS, S. L., SHEARIN, R. L. AND PORTER, C. R. (1971). "A rapid method for determining ^{85}Kr in environmental air samples," page 163 in *Rapid Methods for Measuring Radioactivity in the Environment* (International Atomic Energy Agency, Vienna).

CURRAN, S. C. AND WILSON, H. W. (1965). "Proportional counters and pulse ion chambers," page 303 in *Alpha- Beta- and Gamma-Ray Spectroscopy*, Vol. 1, Siegbahn, K., Ed. (North Holland Publishing Co., Amsterdam).

CURRIE, L. A. (1968). "Limits for qualitative detection and quantitative determination—application to radiochemistry," Anal. Chem. **40**, 586.

DAHL, J. B. AND ØDEGAARD, H. (1970). "Aerial measurements of water equivalent of snow deposits by means of natural radioactivity in the ground," page 191 in *Isotope Hydrology 1970* (International Atomic Energy Agency, Vienna).

DAHLSTROM, T. S. (1965). *Development of a Scintillation Dosimeter System*, Report S-310-R, EGG-1183-2061 (E. G. & G., Inc., Las Vegas, Nevada).

DANIELS, F., BOYD, C. A. AND SAUNDERS, D. F. (1953). "Thermoluminescence as a research tool," Science **117**, 343.

DARNLEY, A. G. (1970). "Airborne gamma-ray spectrometry," CIM Transactions **73**, 20. [Canadian Institute of Mining].

DARNLEY, A. G. AND GRASTY, R. L. (1971). "Mapping from the air by gamma-ray spectrometry," page 495 in *Proc. Third Intl. Geochemical Exploration Symposium*, CIM Special Vol. 11 (Canadian Institute of Mining, Toronto, Canada).

DAVIS, F. J. AND REINHARDT, P. W. (1957). "Instrumentation in aircraft for radiation measurement," Nucl. Sci. Eng. **2**, 713.

DAVIS, F. J. AND REINHARDT, P. W. (1962). *Extended and Point Source Radiometric Program*, Report CEX-60-3 (U. S. Atomic Energy Commission, Washington).

DE, A. K., KHOPKAR, S. M. AND CHALMERS, R. A. (1970). *Solvent Extraction of Metals* (Von Nostrand-Reinhold, New York).

DEAL, L. J., DOYLE, J. F., BURSON, Z. G. AND BOYNS, P. K. (1971). "Locating the lost Athena missile in Mexico by the Aerial Radiological Measuring System (ARMS)," Health Physics **23**, 95.

DEBORTOLI, M. C. (1967). "Radiochemical determination of plutonium in soil and other environmental samples," Anal. Chem. **39**, 375.

DE CAMPO, J. A., BECK, H. L. AND RAFT, P. D. (1972). *High Pressure Argon Ionization Chamber System for the Measurement of Environmental Exposure Rates*, Report HASL-260 (U. S. Atomic Energy Commission, New York).

DEHAAN, A., LEVENTHAL, L. AND BENSON, P. (1965). "A comparative study of the methods used for the computer resolution of composite gamma-ray spectra," page 169 in *Radioisotope Sample Measurement Techniques in Medicine and Biology* (International Atomic Energy Agency, Vienna).

DENHAM, D. H., KATHREN, R. L. AND CORLEY, J. P. (1972). *A CaF_2:Dy Thermoluminescent Dosimeter for Environmental Monitoring*, Report BNWL-SA-4191 (Battelle Northwest Laboratories, Richland, Washington).

DEVOE, J. R. (1962). *Application of Distillation Techniques to Radiochemical Separations*, Report NAS-NS-3108 (National Technical Information

Service, U. S. Department of Commerce, Springfield, Virginia).

DIAMOND, H., FIELDS, P. R., STEVENS, C. S., STUDIER, M. H., FIELD, S. M., INGHRAM, M. G., HESS, D. C., PYLE, G. L., MECH, J. F. AND MANNING, W. M. (1960). "Heavy isotope abundances in Mike thermonuclear device," Phys. Rev. **119**, 2000.

DMITRIEV, A. V., KOGAN, R. M., NIKIFAROV, M. W. AND FRIDMAN, SH. D. (1970a). "Gamma-ray survey of snow cover from aircraft," page 100 in *Meteorologiya i Gidrologiya* No. 3, [Meteorology and Hydrology].

DMITRIEV, A. V., KOGAN, R. M., NAZAROV, I. M., NIKIFAROV, M. V. AND FRIDMAN, SH. D. (1970b). "Procedure for determining the moisture content in the snow cover by gamma radiation of the products of radioactive contamination of the territory," page 88 in *Meterologiya i Gidrologiya* No. 10, [Meteorology and Hydrology].

DMITRIEV, A. V., KOGAN, R. M., NIKIFAROV, M. V. AND FRIDMAN, SH. D. (1971). "Principles of aircraft gamma-ray survey of snow cover," *The Hydro-meteorological Service of the USSR*, (International Union of Geodesy and Geophysics, Toronto, Canada).

DOBOSZ, E. (1973). "A simple method for determination of radioiodine," Radiochem. Radioanal. Letters **13**, 381.

DOIG, R. (1968). "The natural gamma-ray flux: in situ analysis," Geophysics **23**, 311.

DORMAN, L. I. (1972). *Meteorologicheskie Effekti Kosmicheskikh Luchei* [Meteorological Effects of Cosmic Rays] (Izdatel; stvo Nauka, Moscow).

DORMAN, L. I., KAMINER, N. S., KOIAVA, V. K., SHAFER, YU. G. AND SHWARZMAN, B. F. (1956). "Observation of the large cosmic ray increase of February 23, 1956 in the USSR," Nucl. Phys. **1**, 585.

DOUGLAS, G. S., Ed. (1967). *Radioassay Procedures for Environmental Samples,* Public Health Service Publication No. 999-RH-27 (U. S. Department of Health, Education, and Welfare, Washington).

DOYLE, J. F. (1972). "The Aerial Radiological Measuring System (ARMS) Program," page 589 in *The Natural Radiation Environment II*, Adams, J. A. S., Lowder, W. M. and Gesell, T., Eds., Report CONF-720805 (U. S. Energy Research and Development Administration, Washington).

DREVER, R. W. P., MOLJK, A. AND CURRAN, S. C. (1957). "A proportional counter system with small wall effect," Nucl. Instrum. Meth. **1**, 41.

DROBINSKI, J. C., JR., LAGATTA, D. P., GOLDIN, A. S. AND TERRILL, J. G., JR. (1965). "Analyses of environmental samples for carbon-14 and tritium," Health Physics **11**, 385.

DUVAL, J. S., JR., WORDEN, J. M., CLARK, R. B. AND ADAMS, J. A. S. (1972). "Experimental comparison of NaI(Tl) and solid organic scintillation detectors for use in remote sensing of terrestrial gamma rays," Geophysics **37**, 879.

EAKINS, J. D. AND BROWN, D. A. (1966). "Improved method for the simultaneous determination of ^{55}Fe and ^{59}Fe in blood by liquid scintillation counting," Int. J. Appl. Radiat. Isotopes **17**, 391.

EAKINS, J. D. AND GOMM, P. J. (1968). "The determination of gross alpha in urine by adsorption on glass fibre filter paper," Health Physics **14**, 461.

EBERHARDT, L. L. AND GILBERT, R. O. (1972). *Statistical Analysis of Soil Plutonium Studies, Nevada Test Site,* Report BNWL-B-217 (Battelle Northwest Laboratories, Richland, Washington).

EDGINGTON, D. N. AND LUCAS, H. F., JR. (1970). "A system for the neutron activation analysis of trace elements in samples of biological and environmental interest," J. Radioanal. Chem. **5**, 233.

EICHHOLZ, G. G. AND GALLI, A. N. (1970). "Rapid determination of low-level fission products in water," Radiochem. Radioanal. Letters **4**, 315.

EISENBUD, M. (1973). *Environmental Radioactivity,* 2nd ed. (Academic Press, New York).

ELLIS, Y. A. (1970a). "Nuclear data sheets for A = 238," Nucl. Data Sheets **B4**, 581.

ELLIS, Y. A. (1970b). "Nuclear data sheets for A = 234," Nucl. Data Sheets **B4**, 581.

ELLIS, Y. A. (1970c). "Nuclear data sheets for A = 230," Nucl. Data Sheets **B4**, 543.

ELLIS, Y. A. (1971). "Nuclear data sheets for A = 241," Nucl. Data Sheets **B6**, 621.

EMERY, E. W. (1966). "Geiger Müller and proportional counters," page 73 in *Radiation Dosimetry,* Vol. II, Attix, F. H. and Roesch, W. C., Eds. (Academic Press, New York).

EPA (1972). U. S. Environmental Protection Agency, *Environmental Monitoring Surveillance Guide,* Report ORP/SID 72-2 (U. S. Environmental Protection Agency, Washington).

EVANS, R. D. (1969). "Engineer's guide to the elementary behavior of radon daughters," Health Physics **17**, 229.

FAIRMAN, W. D. AND SEDLET, J. (1966). "A rapid and sensitive method for the determination of iodine-131 in milk by ion exchange on silver chloride," Anal. Chem. **38**, 1171.

FAO (1962). Food and Agriculture Organization, *Organization of Surveys for Radionuclides in Food and Agriculture* (United Nations, Rome).

FIELDS, P. R., DIAMOND, H., METTA, D. N., STEVENS, C. M., ROKOP, D. J. AND MORELAND, P. D. (1970). "Isotopic abundances of actinide elements in lunar material," Science **167**, 499.

FIELDS, P. R., DIAMOND, H., METTA, D. N., STEVENS, C. M. AND ROKOP, D. J. (1971). "Isotopic abundances of actinide elements in Apollo 12 samples," page 1571 in *Proc. of Second Lunar Science Conf.* Vol. 2 (The MIT Press, Cambridge, Massachusetts).

FIELDS, P. R., DIAMOND, H., METTA, D. N., ROKOP, D. J. AND STEVENS, C. M. (1972). "^{237}Np, ^{236}U and other actinides on the moon," Geochim. Cosmochim. Acta, Suppl. 3, **2**, 1637.

FITZSIMMONS, C. K., HORN, W. AND KLEIN, W. L. (1972). "A comparison of film badges and thermoluminescent dosimeters used for environmental monitoring," Radiation Data and Reports **13**, 537.

FLEISCHER, R. L. AND LOVETT, D. B. (1968). "Uranium and boron content of water by particle track etching," Geochim. Cosmochim. Acta **32**, 1126.

FLEISCHER, R. L., PRICE, P. B. AND WALKER, R. M. (1965). "Solid state track detectors: applications to nuclear science and geophysics," Ann. Rev. Nucl. Sci. **15**, 1.

FLYNN, W. W. AND MEEHAN, W. R. (1973). "A solvent extraction method for the determination of phosphorus-32 in sea water," Anal. Chim. Acta **63**, 483.

FOLSOM, T. R. AND SREEKUMARAN, C. (1970). "Some reference methods for determining radioactive and natural cesium for marine studies," page 129 in *Reference Methods for Marine Radioactivity Studies* (International Atomic Energy Agency, Vienna).

FOOTE, R. S. (1964). "Time variation of terrestrial γ radiation," page 757 in *The Natural Radiation Environment,* Adams, J. A. S. and Lowder, W. M., Eds. (University of Chicago Press, Chicago).

FOOTE, R. S. (1968). "Review of radiometric techniques in petroleum exploration," page 43 in *Proc. of Symp. on Unconventional Methods in Exploration for Petroleum and Natural Gas,* Report CONF-680985 (U. S. Atomic Energy Commission, Washington).

FOOTE, R. S. (1969). "Improvement in airborne gamma radiation data analysis for anomalous radiation by removal of environmental and pedologic radiation changes," page 187 in *Nuclear Techniques and Mineral Resources* (International Atomic Energy Agency, Vienna).

FORBUSH, S. E. (1958). "Cosmic-ray intensity variation during two solar cycles," J. Geophys. Res. **63**, 651.

FOWLER, E. B., HENDERSON, R. W. AND MILLIGAN, M. F. (1971). *Proceedings of Environmental Plutonium Symposium,* Report LA-4756 (Los Alamos Scientific Laboratory, Los Alamos, New Mexico).

FRIEDLANDER, G., KENNEDY, J. W. AND MILLER, J. M. (1964). *Nuclear and Radiochemistry,* 2nd ed. (John Wiley and Sons, New York).

FRC (1960). Federal Radiation Council, *Background Material for the Development of Radiation Protection Standards,* Report No. 1 (Federal Radiation Council, Washington).

FRITZSCHE, A. E. AND BURSON, Z. G. (1972). "Aerial snow gauging using natural terrestrial gamma radiation," page 607 in *The Natural Radiation Environment II,* Adams, J. A. S., Lowder, W. M. and Gesell, T., Eds., Report CONF-720805 (U. S. Energy Research and Development Administration, Washington).

FRITZSCHE, A. E. AND BURSON, Z. G. (1973). Personal communication (E. G. & G., Inc., Las Vegas, Nevada).

FRITZSCHE, A. E., BURSON, Z. G. AND BURGE, D. L. (1971). *Water Equivalent of Snow Measurements using Natural Terrain Radiation,* Report L-1047, EGG-1183-1533 (E. G. & G., Inc., Las Vegas, Nevada).

FRYER, G. E. AND ADAMS, J. A. S. (1974). "An airborne gamma ray spectrometer and its application in nuclear power plant site surveys," IEEE Trans. on Nucl. Sci., **NS-21**, 572.

GABAY, J. J., PAPERIELLO, C. J., GOODYEAR, S., DALEY, J. C. AND MATUSZEK, J. M. (1974). "A method for determining iodine-129 in milk and water," Health Physics **26**, 89.

GANZERLI-VALENTINI, M. J., MAXIA, V., MELONI, S. AND MARTINELLI, A.

(1971). "Sorption of some actinide ions on ammonium molybdophosphate chromatographic columns," Radiochem. Radioanal. Letters **7**, 163.

GEDEONOV, L. I., ZHILKINA, M. I., BLINOV, V. A., GRITCHENKO, Z. G. AND KUZNETSOV, M. I. (1972). "The temperature system of the calcination of samples of global radioactive fallout," Radiochemiya **14**, 451 (English Translation by Consultants Bureau, New York).

GELSKY, D. E., PLATO, P. AND JACOBSON, A. P. (1970). "In situ measurement of environmental radioactivity," page 278 in *Health Physics Aspects of Nuclear Facility Siting*, Voilleque, P. G. and Baldwin, B. R., Eds. (Health Physics Society, Idaho Falls, Idaho).

GEORGE, M. J. (1970). "New data on the absolute cosmic ray ionization in the lower atmosphere," J. Geophys. Res. **75**, 3693.

GIBSON, J. A. B., RICHARDS, J. E. AND DOGHERTY, J. (1969). "Nuclear radiation in the environment—beta and gamma-ray dose rates and air ionization from 1951 to 1968," J. Atmos. and Terr. Phys. **31**, 1183.

GINGRICH, J. E. AND LOVETT, D. B. (1972). "A track etch technique for uranium exploration," Trans. Am. Nucl. Soc. **15**, 118.

GLEIT, C. E. AND HOLLAND, W. D. (1962). "Use of electrically excited oxygen for the low temperature decomposition of organic substances," Anal. Chem. **34**, 1454.

GODEBOLD, B. C. AND JONES, J. K., Eds. (1965). *Radiological Monitoring of the Environment* (Pergamon Press, Oxford).

GOLCHERT, N. W. AND SEDLET, J. (1969). "Radiochemical determination of technetium-99 in environmental water samples," Anal. Chem. **41**, 669.

GOLD, S., BARKLAU, H. W., SHLEIEN, B. AND KAHN, B. (1964). "Measurement of naturally occurring radionuclides in air," page 369 in *The Natural Radiation Environment*, Adams, J. A. S. and Lowder, W. M., Eds. (University of Chicago Press, Chicago).

GOLD, R., STRASH, A., CONGEL, F. J. AND ROBERTS, J. H. (1973). "Continuous gamma ray spectroscopy in the natural environment," IEEE Trans. on Nucl. Sci. **NS-20**, 48.

GOLD, R., OLDMAN, B. G., ECKERMAN, K. G. AND STRASH, A. M. (1974). "Environmental radiation at the EBR-II site," IEEE Trans. on Nucl. Sci. **NS-21**, 596.

GORSUCH, T. T. (1970). *The Destruction of Organic Matter* (Pergamon Press, Oxford).

GOUSSEV, N. G., KOVALEV, E. E. AND FODERARO, H. (1968). "Derivation of formulae," page 367 in *Engineering Compendium on Radiation Shielding*, Vol. I (Springer-Verlag, New York).

GRASTY, R. L. (1972). Personal Communication (Geological Survey of Canada, Ottawa).

GRASTY, R. L. AND DARNLEY, A. G. (1971). "The calibration of gamma-ray spectrometers for ground and airborne use," Research paper 71-17 (Geological Survey of Canada, Ottawa, Canada).

GRAY, L. H. (1936). "An ionization method for the absolute measurement of gamma-ray energy," Proc. Roy. Soc. **A 156**, 578.

GREGORY, A. F. AND HORWOOD, J. L. (1963). *A Spectrometric Study of the Attenuation in Air of Gamma Rays from Mineral Sources,* Report R110 (Geological Survey of Canada, Ottawa, Canada).

GREVET, P. (1966). "Mesure de l'intensité totale d'ionization au voisinage du sol et separation de ses differentes composantes," Tellus **18,** 587.

GROBUSHINA, L. B., ZHILTSOVA, L. YA., MATVEVVA, E. N., SURGANOVA, N. A., TENYALV, V. G. AND TYMINSKY, V. G. (1972). "Extraction-scintillation method for determining the concentration of radioisotopes," J. Radioanal. Chem. **10,** 165.

GRODZINSKII, D. M. (1965). *Estestvennaya Radioaktivnost' Rastenii i Pochv* [Natural Radioactivity of Plants and Soils] (Naukova Dumka, Kiev).

GROTCH, H. (1962). *Energy Absorption Spectrum in Air of Cosmic-Ray Mu Mesons at Sea Level,* Report HASL-130 (U. S. Atomic Energy Commission, New York).

GUILLOU, R. B. (1964). "The Aerial Radiological Measuring Systems (ARMS) program," page 705 in *The Natural Radiation Environment,* Adams, J. A. S. and Lowder, W. M., Eds. (University of Chicago Press, Chicago).

GUNNINK, R. AND MORROW, R. J. (1971). *Gamma-Ray Energies and Absolute Branching Intensities for* $^{238,239,240,241}Pu$ *and* ^{241}Am*,* Report UCRL-51087 (Lawrence Livermore Laboratory, Livermore, California).

GUPTA, G. N. (1966). "A simple in-vial combustion method for assay of hydrogen-3, carbon-14, and sulfur-35 in biological, biochemical and organic materials," Anal. Chem. **38,** 1356.

GUSTAFSON, P. F. AND BRAR, S. (1964). "Measurement of γ-emitting radionuclides in soil and calculation of the dose arising therefrom," page 499 in *The Natural Radiation Environment,* Adams, J. A. S. and Lowder, W. M., eds. (University of Chicago Press, Chicago).

GUTHRIE, J. E. AND GRUMMITT, W. E. (1963). *Manual of Low Activity Environmental Analyses,* Report AECL-1745 (Atomic Energy of Canada, Limited, Chalk River, Ontario).

HAISSINSKY, M. (1964). *Nuclear Chemistry and Its Applications,* translated by Tuck, D. G. (Addison-Wesley, Reading, Massachusetts).

HAMILTON, E. L. (1971). "The relative radioactivity of building materials," Am. Indust. Hyg. Assoc. J. **32,** 398.

HAND, J. E. AND WEISSMAN, V. F. (1968). *ARMS Aircraft Recov[illegible] of Lost Cobalt-60 Source,* Report L-901, EGG-1183-1457 (E. G. & G., Inc., Las Vegas, Nevada).

HARDY, E. P., Ed. (1974). "Strontium-90 in New York City and San Francisco diets," page II-4 in *Fallout Program Quarterly Summary Report,* Report HASL-286 (U. S. Atomic Energy Commission, New York).

HARDY, E. P., KREY, P. W. AND VOLCHOK, H. L. (1973). "Global inventory and distribution of fallout plutonium," Nature **241,** 444.

HARLEY, J. H., Ed. (1972). *HASL Procedures Manual,* Report HASL-300 (U. S. Atomic Energy Commission, New York).

HEALY, J. W. (1971). "Problems of plutonium measurement in the environment," IEEE Trans. on Nucl. Sci. **NS-19,** 219.

HEATH, R. L. (1964). "Scintillation spectrometry gamma ray spectrum catalog," Report IDO-16408 (U. S. Atomic Energy Commission, Idaho Falls, Idaho).

HEATH, R. L. (1969). "Gamma-ray spectrometry and automated data systems for activation analysis," page 1007 in *Modern Trends in Activation Analysis,* Vol. 2, DeVoe, J. R., Ed., National Bureau of Standards Special Publication 321 (National Bureau of Standards, Washington).

HELMER, R. G. AND PUTNAM, M. H. (1972). *Gauss V a Computer Program for the Analysis of Gamma-Ray Spectra from Ge(Li) Spectrometers,* Report ANCR-1043 (Aerojet-Nuclear, Idaho Falls, Idaho).

HENDRICKSON, W. S. AND RIEL, G. K. (1975). "Comparison of ferrocyanides for cesium extraction," Health Physics **28,** 17.

HESS, V. F. AND VANCOUR, R. P. (1950). "The ionization balance of the atmosphere," J. Atmos. Terr. Phys. **1,** 13.

HESS, V. F. AND O'DONNELL, G. A. (1951). "On the rate of ion formation at ground level and at one meter above ground," J. Geophys. Res. **56,** 557.

HESS, V. F. AND MANNING, A. W. (1956). "A study of the ionization produced in various gases," Trans. Am. Geophys. Union **37,** 676.

HESS, V. F., PARKINSON, W. D. AND MIRANDA, H. A. (1953). *Beta Ray Ionization from the Ground,* Fordham University Department of Physics Scientific Report No. 4 (Fordham University, New York).

HIEBERT, R. D., JOHNSON, L. AND KOELLE, A. R. (1973). *Los Alamos Pulse-Height Analyzer,* Report LA-5203-MS (Los Alamos Scientific Laboratory, Los Alamos, New Mexico).

HILL, C. R. (1961). "A method of alpha particle spectroscopy for materials of very low specific activity," Nucl. Instrum. Meth. **12,** 299.

HILLEBRAND, W. F., LUNDELL, G. E. F., BRIGHT, H. A. AND HOFFMAN, J. I. (1953). *Applied Inorganic Analysis* (John Wiley and Sons, New York).

HODGE, V. F. AND FOLSOM, T. R. (1970). "Cyanide extraction and electrodeposition of trace amounts of radioactive silver from large biological samples," Anal. Chem. **44,** 381.

HOFFMAN, D. C., LAWRENCE, F. O., MEWHERTER, J. L. AND ROURKE, F. M. (1971). "Detection of plutonium-244 in nature," Nature **234,** 133.

HOLCOMB, H. D. (1964). "Separation of americium from curium with calcium fluoride," Anal. Chem. **36,** 2329.

HOLT, P. D. AND GIBSON, J. A. B. (1964). *High Energy Gamma-Ray Dosimetry: Experimental Measurements at a Nuclear Power Station,* Report AERE-R 4451 (U. K. Atomic Energy Authority, Harwell, England).

HOLTZMAN, R. B. (1963). "Measurement of the natural contents of RaD (Pb^{210}) and RaF (Po^{210}) in human bone—estimates of whole body burdens," Health Physics **9,** 385.

HONSTEAD, J. F. (1969). "Mechanisms of transfer of radionuclides to man in the vicinity of a nuclear facility," page 381 in *Environmental Contamination by Radioactive Materials* (International Atomic Energy Agency, Vienna).

HONSTEAD, J. F. (1971). "Quantitative evaluation of environmental factors affecting population exposure near Hanford," page 266 in *Health Physics Aspects of Nuclear Facility Siting,* Voilleque, P. G. and Baldwin, B. R.,

Eds. (Health Physics Society, Idaho Falls, Idaho).

HOOGENBOOM, A. M. (1958). "A new method in gamma-ray spectroscopy: a two crystal scintillation spectrometer with improved resolution," Nucl. Instrum. Meth. **3**, 57.

HOPFIELD, J. J. (1933). "Use of argon in the ionization method of measuring cosmic rays and gamma rays," Phys. Rev. **43**, 675.

HOREN, D. J. (1972). "Nuclear data sheets for A = 129," Nucl. Data Sheets **B8**, 123.

HORROCKS, D. L. (1973). "Energy calibration of a liquid scintillation counter," Int. J. Appl. Radiat. and Isotopes **24**, 49.

HORROCKS, D. L. AND STUDIER, M. H. (1964). "Determination of radioactive noble gases with a liquid scintillator," Anal. Chem. **36**, 2077.

HOUTERMANS, F. G. AND OESCHGER, H. (1958). "Proportionzählrohr zur Messung schwacher aktivitäten weicher β-strahlung," Helv. Phys. Acta **31**, 117.

HORWITZ, W., Ed. (1965). *Official Methods of Analysis of the Association of Official Agricultural Chemists* (Association of Official Agricultural Chemists, Washington).

HORWITZ, E. P. AND BLOOMQUIST, C. A. A. (1973). "High speed-high efficiency separation of the transplutonium elements by extraction chromatography," J. Inorg. Nucl. Chem. **35**, 371.

HULTQVIST, B. (1956). *Studies on Naturally Occurring Ionizing Radiations*, Kgl. Svenska Vetenskaps. Handl. Vol. 6, Ser. 4, No. 3 (Almqvist and Wiksells Boktryckeri AB, Stockholm).

IAEA (1961). International Atomic Energy Agency, *Radioactive Waste Disposal into the Sea* (International Atomic Energy Agency, Vienna).

IAEA (1966a). *Disposal of Radioactive Wastes into Seas, Oceans and Surface Waters* (International Atomic Energy Agency, Vienna).

IAEA (1966b). International Atomic Energy Agency, *Manual of Environmental Monitoring in Normal Operation* (International Atomic Energy Agency, Vienna).

IAEA (1969a). International Atomic Energy Agency, *Environmental Contamination by Radioactive Materials* (International Atomic Energy Agency, Vienna).

IAEA (1969b). International Atomic Energy Agency, *Nuclear Techniques and Mineral Resources* (International Atomic Energy Agency, Vienna).

IAEA (1971a). International Atomic Energy Agency, *Disposal of Radioactive Wastes into Rivers, Lakes and Estuaries* (International Atomic Energy Agency, Vienna).

IAEA (1971b). International Atomic Energy Agency, *Environmental Aspects of Nuclear Power Stations* (International Atomic Energy Agency, Vienna).

IAEA (1971c). International Atomic Energy Agency, *Reference Methods for Marine Radioactive Studies* (International Atomic Energy Agency, Vienna).

ICRP (1959). International Commission on Radiological Protection, *Report of Committee II on Permissible Dose for Internal Radiation*, ICRP Publication 2 (Pergamon Press, London).

ICRP (1965a). International Commission on Radiological Protection, *Princi-*

ples of Environmental Monitoring Related to the Handling of Radioactive Materials, ICRP Publication 7 (Pergamon Press, London).

ICRP (1965b). International Commission on Radiological Protection, *Recommendations of the International Commission on Radiological Protection,* ICRP Publication 9 (Pergamon Press, London).

ICRP (1966). International Commission on Radiological Protection, "Radiological aspects of the supersonic transport," Health Physics **12,** 209.

ICRU (1963). International Commission on Radiological Units and Measurements, *Radioactivity,* ICRU Report 10c, published as National Bureau of Standards Handbook 86 (U. S. Government Printing Office, Washington).

ICRU (1964). International Commission on Radiological Units and Measurements, *Physical Aspects of Irradiation,* ICRU Report 10b, published as National Bureau of Standards Handbook 85 (U. S. Government Printing Office, Washington).

ICRU (1971a). International Commission on Radiation Units and Measurements, *Radiation Quantities and Units,* ICRU Report 19 (International Commission on Radiation Units and Measurements, Washington).

ICRU (1971b). International Commission on Radiation Units and Measurements, *Radiation Protection Instrumentation and its Application,* ICRU Report 20 (International Commission on Radiation Units and Measurements, Washington).

ICRU (1972). International Commission on Radiation Units and Measurements, *Measurement of Low-Level Radioactivity,* ICRU Report 22 (International Commission on Radiation Units and Measurements, Washington).

Iida, T. and Kawano, M. (1972). "Measurements of the ionization intensity due to beta radiation near the ground," Arch. Meteor. Geophys. Biokl. Series A, **21,** 195.

Ikebe, Y. (1970). "Evaluation of the total ionization in the lower atmosphere," J. Earth Sci. **18,** 85.

Irving, H. and Edgington, D. N. (1960). "Synergic effects in the solvent extraction of the actinides—I. Uranium (VI)," J. Inorg. Nucl. Chem. **15,** 158.

Israel, H. I., Lier, D. W. and Storm, E. (1971). "Comparison of detectors used in measurement of 10 to 300 keV x-ray spectra," Nucl. Instrum. Meth. **91,** 141.

Israelsson, S., Knudsen, E. and Ungethüm, E. (1972). *Randon and Thoron in the Air Near the Ground,* Marsta Observatory Report No. 28 (Meteorologiska Inst. vid Kungl. Univ., Uppsala, Sweden).

Jacobi, W. (1963). "Die natürliche Radioaktivität der Atmosphäre," Biophysik **1,** 175.

Jacobi, W. and Andre, K. (1963). "The vertical distribution of radon 222, radon 220 and their decay products in the atmosphere," J. Geophys. Res. **68,** 3799.

Jacobs, D. G. (1968). *Sources of Tritium and Its Behavior Upon Release to the Environment,* Report TID-24635 (U. S. Atomic Energy Commission, Washington).

Jantsch, K. (1967). "Triton reactions with ^{30}Si. Part 3. Determination of the

^{32}Si half-life," Kernenergie **10,** 89.

Jaquish, R. E. and Johns, F. B. (1972). "Concentration of krypton-85 in air," page 833 in *The Natural Radiation Environment II,* Adams, J. A. S., Lowder, W. M. and Gesell, T., Eds., Report CONF-720805 (U. S. Energy Research and Development Administration, Washington).

Jaquish, R. E. and Moghissi, A. A. (1973). "Survey of analytical methods for environmental monitoring of krypton-85," page 169 in *Noble Gases,* Stanley, R. E. and Moghissi, A. A., Eds., Report CONF-730915 (U. S. Environmental Protection Agency, Las Vegas, Nevada).

Jaworowski, Z., Bilkiewicz, J., Kownacka, L. and Woldek, S. (1972). "The artificial sources of natural radionuclides in the environment," page 809 in *The Natural Radiation Environment II,* Adams, J. A. S., Lowder, W. M. and Gesell, T., Eds., Report CONF-720805 (U. S. Energy Research and Development Administration, Washington).

Johansson, S. A. E. (1950). "A scintillation spectrometer for high-energy gamma-rays," Nature **166,** 794.

Johns, F. B. (1970). *Southwestern Radiological Health Laboratory Handbook of Radiochemical Analytical Methods,* Report SWRHL-11 (U. S. Department of Health, Education and Welfare, Las Vegas, Nevada).

Johns, F. B. (1973). "Portable apparatus and procedure for the separation of krypton, xenon, and methane in air," page 225 in *Noble Gases,* Stanley, R. E. and Moghissi, A. A., Eds., Report CONF-730915 (U. S. Environmental Protection Agency, Las Vegas, Nevada).

Johnson, L. J. (1972). *Los Alamos Land Areas Environmental Radiation Survey,* Report LA-5097-MS (Los Alamos Scientific Laboratory, Los Alamos, New Mexico).

Jokipii, J. R. (1971). "Propagation of cosmic-rays in the solar winds," Rev. Geophys. Space Sci. **9,** 27.

Jones, A. R. (1966). "Proposed calibration factors for various dosimeters at different energies," Health Physics **12,** 663.

Jones, A. R. (1974). "Measurement of low-level environmental γ doses with TLDs and Geiger counters," IEEE Trans. on Nucl. Sci. **NS-21,** 456.

Kahn, B. (1965). "Determination of picocurie concentrations of iodine-131 in milk," Agric. Food Chem. **13,** 21.

Kahn, B. (1973). "Determination of radioactive nuclides in water," page 1357 in *Water and Water Pollution Handbook,* Vol. 4, Ciaccio, L. L., Ed. (M. Decker, New York).

Kahn, B., Blanchard, R. L., Kolde, A. E., Krieger, H. L., Gold, S., Brinck, W. L., Averett, W. J., Smith, D. B. and Martin, A. (1971). *Radiological Surveillance Studies at a Pressurized Water Nuclear Power Reactor,* Report RD 71-1 (U. S. Environmental Protection Agency, Cincinnati, Ohio).

Kahn, B., Shleien, B. and Weaver, C. (1972). "Environmental experience with radioactive effluents from operating nuclear power plants," page 559 in *Peaceful Uses of Atomic Energy,* Vol. 11 (United Nations, New York).

Kanapilly, G. M. and Yaeger, D. J. (1969). "Colloid filtration as a preparative method for the radiochemical analysis of biological samples," Health Physics **17,** 579.

KARCHES, G. J., KOLDE, H. E., BRINCK, W. L., SHEARIN, R. L. AND PHILLIPS, C. R. (1971). "Field determination of dose from ^{133}Xe in the plume from a pressurized water reactor," page 515 in *Rapid Methods for Measuring Radioactivity in the Environment* (International Atomic Energy Agency, Vienna).

KASTNER, J., SHONKA, F. R. AND ROSE, J. E. (1964). "A muscle-equivalent environmental radiation meter of extreme sensitivity," page 655 in *The Natural Radiation Environment,* Adams, J. A. S. and Lowder, W. M., Eds. (University of Chicago Press, Chicago).

KAWANO, M., IKEBE, Y., NAKASHIMA, Y. AND SHIMIZU, K. (1965). "Some properties of naturally occurring radiations and radioactive ions in the atmosphere," Memoirs Faculty Engineering, Nagoya Univ. **17**, 125.

KAWANO, M., IKEBE, Y., NAKASHIMA, Y., SEKIGUCHI, N. AND NAKAYAMA, T. (1969). "The ionization of the air due to beta- and gamma-radiations near the ground, I and II," J. Radiat. Res. (Japan) **10**, 139, 145.

KIRK, W. P. (1972). *Krypton-85, A Review of the Literature and an Analysis of Radiation Hazards,* Report PB-207079 (U. S. Environmental Protection Agency, Montgomery, Alabama).

KLECHKOVSKII, V. M., POLIKARPOV, G. G. AND ALEKSAKHIN, R. M., Eds. (1973). *Radioecology* (Halsted Press, New York).

KLEMENT, A. W., JR. (1965). *Natural Environmental Radioactivity, An Annotated Bibliography,* Report WASH-1061 (U. S. Atomic Energy Commission, Washington).

KNOX, J. B. AND PETERSON, K. R. (1972). "Estimates of dose to northern hemisphere from ^{85}Kr emitted by a single nuclear fuel-reprocessing plant," Nucl. Safety **13**, 130.

KOCH, R. C. AND GRANDY, G. L. (1957). *Retention Efficiencies of Charcoal Traps for Fission Gases,* Report NSEC-1 (Nuclear Science and Engineering Corp., Pittsburgh, Pennsylvania).

KOGAN, R. M. AND FRIDMAN, SH. D. (1967). "Effect of rock and soil moisture on the structure of the gamma field in the near-surface atmosphere," Earth Physics No. **3**, 82.

KOGAN, R. M., NAZAROV, I. M. AND FRIDMAN, SH. D. (1969). *Gamma Spectrometry of Natural Environments and Formations* (Atomizdat, Moscow) [English translation: Israel Program for Scientific Translations, Jerusalem, available from National Technical Information Service, Springfield, Virginia].

KOL', S. A. (1952). "Investigation of the infiltration capacity of soils of agricultural fields and forest belts in Kamennaya Steppe," Trans. State Hydrological Institut **34**, 88.

KOLB, W. AND LAUTERBACH, U. (1972). "A scintillation dose rate meter for environmental radiation surveys," page 245 in *The Natural Radiation Environment II,* Adams, J. A. S., Lowder, W. M. and Gesell, T., Eds., Report CONF-720805 (U. S. Energy Research and Development Administration, Washington).

KOLTHOFF, I. M. AND ELVING, P. J. (1959 to present). *Treatise on Analytical Chemistry* (Wiley-Interscience, New York).

KORENMAN, I. (1966). *Analytical Chemistry of Low Concentrations,* Translated from Russian by J. Schmarak (Israel Program for Scientific Translations, Jerusalem, 1968).

KORTHOVEN, P. J. M. (1972). *ICPEAX, A Computer Program for the Automatic Analysis of Gamma Ray Spectra Obtained with High Resolution Semi-conductor Detectors,* Report IRI 133-70-4 (Inter-universitrair Reactor Instituut, Delft, Netherlands).

KRAUS, K. A. AND NELSON, F. (1957). "Radiochemical separations by ion exchange," Ann. Rev. Nucl. Sci. **7,** 31.

KREY, P. W. AND HARDY, E. P. (1970). *Plutonium in Soil Around the Rocky Flats Plant,* Report HASL-235 (U. S. Atomic Energy Commission, New York).

KRIEGER, H. L. AND GOLD, S. (1973). *Procedures for Radiochemical Analysis of Nuclear Reactor Aqueous Solutions,* Report EPA-R4-73-014 (U. S. Environmental Protection Agency, Cincinnati, Ohio).

KRIEGER, H. L., VELTEN, R. J. AND BURMANN, F. J., Eds. (1966). *Radionuclide Analysis of Environmental Samples,* Robert A. Taft Sanitary Engineering Center Report R-59-6 (U. S. Department of Health, Education and Welfare, Cincinnati, Ohio).

KUPER, J. B. H. AND CHASE, R. L. (1950). "Monitor for low intensity gamma rays," Rev. Sci. Instrum. **21,** 356.

LAL, D. AND PETERS, B. (1967). "Cosmic ray produced radioactivity on the earth," page 551 in *Encyclopedia of Physics,* Vol. 46, No. 2, Cosmic Rays II, Sitte, K., Ed. (Springer-Verlag, Berlin).

LAL, D., ARNOLD, J. R. AND SOMAYAJULU, B. L. K. (1964). "A method for the extraction of trace elements from sea water," Geochim. Cosmochim. Acta **28,** 1111.

LAMANNA, A. YANCIK, V., ABBOT, T. P. AND HOLMES, J. (1965). "Thimerosal, a preservative for milk to be tested for radioactivity," Health Physics **11,** 586.

LARSEN, R. P. AND OLDHAM, R. D. (1975). "Anion-exchange separation of plutonium in hydrochloric-hydrobromic acid media," Talanta, 22, 577.

LAVRUKHINA, A. K., MOLYSHEVA, T. V. AND PAVLOTSKAYA, F. I. (1967). *Chemical Analysis of Radioactive Materials* (Chemical Rubber Co., Cleveland, Ohio) [English Translation of Radiokhimicheskii analiz, USSR Academy of Sciences].

LEDERER, C. M., HOLLANDER, J. M. AND PERLMAN, I. (1967). *Table of Isotopes* (John Wiley and Sons, New York).

LEMLICH, R. (1972). *Absorptive Bubble Separation Techniques* (Academic Press, New York).

LENGEMANN, F. W. (1966). "Predicting the total projected intake of radioiodine from milk by man—I, the situation where no countermeasures are taken," Health Physics **12,** 825. "II, the situation where countermeasures are taken," Health Physics **12,** 831.

LENTSCH, J. W., WRENN, M. E., KNEIP, T. J. AND EISENBUD, M. (1971). "Manmade radionuclides in the Hudson River estuary," page 499 in *Health Physics Aspects of Nuclear Facility Siting,* Voilleque, P. G. and Baldwin, B. R., Eds. (Health Physics Society, Idaho Falls, Idaho).

LEVIN, S. G. AND STOMS, R. K. (1969). "Natural background gamma radiation dose rate measurements in Michigan, Colorado and Minnesota," Am. J. Public Health **59**, 102.

LEWIS, M. B. (1971a). "Nuclear data sheets for A = 208," Nucl. Data Sheets **B5**, 243.

LEWIS, M. B. (1971b). "Nuclear data sheets for A = 210," Nucl. Data Sheets **B5**, 631.

LIBBY, W. F. (1955). "Dosages from natural radioactivity and cosmic rays," Science **123**, 57.

LIBBY, W. F. (1957). "Simple absolute measurement technique for beta radioactivity," Anal. Chem. **29**, 1567.

LIBOFF, A. R. (1972). "Cosmic ray ionization in the lower atmosphere," page 55 in *The Natural Radiation Environment II,* Adams, J. A. S., Lowder, W. M. and Gesell, T., Eds., Report CONF-720805 (U. S. Energy Research and Development Administration, Washington).

LINDEKEN, C. L. (1961). "Use of natural airborne radioactivity to evaluate filters for alpha air sampling," Am. Ind. Hyg. Assoc. J. **22**, 232.

LINDEKEN, C. L. (1973). Personal communication (Lawrence Livermore Laboratory, Livermore, California).

LINDEKEN, C. L. AND HODGKINS, D. J. (1969). *Response of the FIDLER Detector to Low Energy Natural Background Radiation in the Livermore Valley,* Report UCRL-50007-69-3 (Lawrence Livermore Laboratory, Livermore, California).

LINDEKEN, C. L., PETERSON, K. R., JONES, D. E. AND MCMILLEN, R. E. (1972). "Geographical variations in environmental radiation background in the United States," page 317 in *The Natural Radiation Environment II,* Adams, J. A. S., Lowder, W. M. and Gesell, T., Eds., Report CONF-720805 (U. S. Energy Research and Development Administration, Washington).

LINDENBAUM, A. AND WESTFALL, W. (1965). "Colloidal properties of plutonium in dilute solution," Int. J. Appl. Radiat. and Isotopes **16**, 545.

LIPPERT, J. AND MEJDAHL, V. (1967). "Thermoluminescent readout instrument for measurement of small doses," page 204 in *Luminescence Dosimetry,* Symposium Series 8, Attix, F. H., Ed. (U. S. Atomic Energy Commission, Washington).

LLACER, J. (1972). "Planar and coaxial high purity germanium radiation detectors," Nucl. Instrum. Meth. **98**, 259.

LOCKHART, L. B., JR. (1964). "Radioactivity of the radon-222 and radon-220 series in the air at ground level," page 331 in *The Natural Radiation Environment,* Adams, J. A. S. and Lowder, W. M., Eds. (University of Chicago Press, Chicago).

LOCKHART, L. B., JR. AND PATTERSON, R. L., JR. (1962). "Intercalibration of some air monitoring systems," Radiol. Health Data **3**, 468.

LOCKHART, L. B., JR. AND PATTERSON, R. L., JR. (1964). "Intercalibration of the major North American monitoring system," Radiol. Health Data **5**, 12.

LOCKHART, L. B., JR., PATTERSON, R. L., JR. AND ANDERSON, W. L. (1964). *Characteristics of Air Filter Media Used for Monitoring Airborne Radioactivity,* Report NRL-6054 (Naval Research Laboratory, Washington).

LOCKWOOD, J. A. (1971). "Forbush decreases in the cosmic radiation," Space Sci. Rev. **12,** 568.

LOCKWOOD, J. A., LEZNIAK, J. A. AND WEBBER, W. R. (1972). "Change in the eleven-year modulation at the time of the June 8, 1969 Forbush decrease," J. Geophys. Res. **77,** 4839.

LÖVBORG, L., KUNZENDORF, H. AND HANSEN. J. (1969). "Use of field gamma-spectrometry in the exploration of uranium and thorium deposits in south Greenland," page 197 in *Nuclear Techniques and Mineral Resources* (International Atomic Energy Agency, Vienna).

LOWDER, W. M. (1969). *Low-Energy Atmospheric Gamma Radiation as a Measure of Cosmic-Ray Secondary Electron Flux,* Report HASL-209 (U. S. Atomic Energy Commission, New York).

LOWDER, W. M. AND SOLON, L. R. (1956). *Background Radiation, A Literature Search,* Report NYO-4712 (U. S. Atomic Energy Commission, New York).

LOWDER, W. M. AND CONDON, W. J. (1965). "Measurement of the exposure of human populations to environmental radiation," Nature **206,** 658.

LOWDER, W. M. AND BECK, H. L. (1966). "Cosmic-ray ionization in the lower atmosphere," J. Geophys. Res. **71,** 4661.

LOWDER, W. M. AND O'BRIEN, K. (1972). *Cosmic-Ray Dose Rates in the Atmosphere,* Report HASL-254 (U. S. Atomic Energy Commission, New York).

LOWDER, W. M., RAFT, P. D. AND BECK, H. L. (1971). "Experimental determinations of cosmic-ray charged particle intensity profiles in the atmosphere," page 908 in *Natural and Manmade Radiation in Space,* Warman, E. A., Ed., Report NASA TM-X-2440 (National Aeronautics and Space Administration, Washington).

LOWDER, W. M., BECK, H. L., RAFT, P. D., BURKE, G. DE P., GOGOLAK, C. V., DECAMPO, J. A. AND MCLAUGHLIN, J. E. (1972). "Experimental investigations of the environmental radiation field," page 255 in *The Natural Radiation Environment II,* Adams, J. A. S., Lowder, W. M. and Gesell, T., Eds., Report CONF-720805 (U. S. Energy Research and Development Administration, Washington).

LOWDER, W. M., RAFT, P. D. AND GOGOLAK, C. V. (1973). *Environmental Gamma Radiation from Nitrogen-16 Decay in the Turbines of a Large Boiling Water Reactor,* Report HASL-271 (U. S. Atomic Energy Commission, New York).

LUCAS, H. F. (1957). "Improved low-level alpha scintillation counter for radon," Rev. Sci. Instrum. **28,** 680.

LUCAS, H. F., JR., EDGINGTON, D. W. AND MARKIN, F. (1970). "Natural thorium in human bone," Health Physics **19,** 739.

MADIGOSKY, S. W. AND SIMONS, D. G. (1966). *A Monte Carlo Calculation to Simulate Detector Response to Underwater Gamma Rays,* Report NOLTR 66-140 (Naval Surface Weapons Center, Silver Spring, Maryland).

MAHDAVI, A. (1964). "The thorium, uranium and potassium contents of Atlantic and Gulf Coast beach sands," page 87 in *The Natural Radiation Environment,* Adams, J. A. S. and Lowder, W. M., Eds. (University of Chicago Press, Chicago).

MAJOR, W. J., WESSMAN, R. A., MELGARD, R. AND LEVENTHAL, L. (1964). "Routine determination of plutonium by tracer techniques in large biological samples," Health Physics **10**, 957.

MALAKHOV, S. G. AND CHERNYSHEVA, P. G. (1965). "On the seasonal variations in the concentration of radon and thoron in the surface layer of the atmosphere," page 60 in *Radioactive Isotopes in the Atmosphere and Their Use in Meteorology*, Karol', I. L., Kirichenko, L. V., Krasnopevtsev, Yu. V., Kurganskaya, V. M., Malakhov, S. G., Sereda, G. A., and Yagodovskii, I.V. Eds. (Atomizdat, Moscow) [English translation by Israel Program for Scientific Translations, Jerusalem, 1967].

MARCUS, Y. AND KERTES, A. S. (1969). *Ion Exchange and Solvent Extraction of Metal Complexes* (John Wiley and Sons, New York).

MARTIN, A. AND BLANCHARD, R. L. (1969). "The thermal volatilization of caesium-137, polonium-210, and lead-210 from in vivo labelled samples," Analyst **94**, 441.

MARTIN, J. E., HARWARD, E. D., OAKLEY, D. T., SMITH, J. M. AND BEDROSSIAN, P. H. (1971). "Radioactivity from fossil-fuel and nuclear power plants," page 325 in *Environmental Aspects of Nuclear Power Stations* (International Atomic Energy Agency, Vienna).

MARTIN, M. J. (1973). *Radioactive Atoms – Supplement 1*, Report ORNL-4923 (Oak Ridge National Laboratory, Oak Ridge, Tennessee).

MARTIN, M. J. AND BLICHERT-TOFT, P. H. (1970). "Radioactive atoms," Nucl. Data Tables **A8**, 1.

MARTZ, D. E., HOLLEMAN, D. E., MCCURDY, D. E. AND SCHAIGER, K. J. (1969). "Analysis of atmospheric concentrations of RaA, RaB, and RaC by alpha spectroscopy," Health Physics **17**, 131.

MASSART, D. L. (1971). *Cation-Exchange Techniques in Radiochemistry*, Report NAS-NS-3113 (National Technical Information Service, U. S. Department of Commerce, Springfield, Virginia).

MATHEWS, T., QUENBY, J. AND SEAR, N. (1971). "Mechanism for cosmic ray modulation," Nature **229**, 246.

MCCASLIN, J. B. (1964). "Electrometer for ionization chambers using metal-oxide-semiconductor field-effect transistors," Rev. Sci. Instrum. **35**, 1587.

MCDOWELL, W. J., FARRAR, D. T. AND BILLINGS, M. R. (1974). "Plutonium and uranium determinations in environmental samples: combines solvent extraction – liquid scintillation method," Talanta **21**, 1231.

MCHENRY, J. R. AND GILL, A. C. (1970). "Measurement of soil moisture with a portable gamma-ray scintillation spectrometer," Water Resources Research **6**, 989.

MCHUGH, J. A. (1969). "Surface ionization-the rhenium v-type single filament," Int. J. Mass. Spectrom. Ion Phys. **3**, 267.

MCKAY, H. A. C. (1971). *Principles of Radiochemistry* (Butterworths, London).

MCKAY, H. A. C., HEALY, T. V., JENKINS, I. C. AND NAYLOR, A., Eds. (1965). *Solvent Extraction Chemistry of Metals* (MacMillan, London).

MCLAUGHLIN, J. E. AND BECK, H. L. (1973). "Environmental radiation dosimetry for nuclear facilities and problems," IEEE Trans. on Nucl. Sci. **NS-20**, 36.

MEDSKER, L. R. AND HOREN, D. J. (1972). "Nuclear data sheets for A = 95," Nucl. Data Sheets **B8**, 29.

MEGAW, W. J. AND MAY, F. G. (1962). "The behavior of iodine released in reactor containers," Reactor Sci. Tech. **16**, 427.

MEJDAHL, V., Ed. (1971). *Proc. of Third Intl. Conf. on Luminescence Dosimetry,* Rpt. 249 (Danish AEC Research Establishment, Risø, Denmark).

MENON, M. P., MENON, K. K. AND KURODA, P. (1963). "On the stratospheric fallout of bomb produced cerium isotopes," J. Geophys. Res. **68**, 4495.

MERIAN, R. F., LACKEY, J. G. AND HAND, J. E. (1960). *Aerial Radiological Monitoring System, Theoretical Analysis, Design and Operation of a Revised System,* Vol. I, Report CEX-59.4 (E. G. & G., Inc., Las Vegas, Nevada).

MILLER, J. M. AND SYMONS, G. D. (1973). "Radiometric traverse of the sea bed off the Yorkshire coast," Nature **242**, 184.

MILLIKAN, R. A. (1932). "Cosmic-ray ionization and electroscope constants as a function of pressure," Phys. Rev. **39**, 297.

MILLIKAN, R. A. AND NEHER, H. V. (1936). "A precision world survey of sea-level cosmic-ray intensities," Phys. Res. **50**, 15.

MINATO, S. (1971). "Terrestrial gamma radiation field in natural environment," J. Nucl. Sci. Tech. **8**, 342.

MINATO, S. AND KAWANO, M. (1970). "On the constitution of terrestrial gamma radiation," J. Geophys. Rev. **75**, 5825.

MIRANDA, H. A. (1958). "The ionization due to beta radiation from the atmosphere," J. Geophys. Res. **63**, 147.

MITCHELL, R. F. (1960). "Electrodeposition of actinide elements at tracer concentrations," Anal. Chem. **32**, 327.

MOGHISSI, A. A., MCNELIS, D. N., PLATT, W. F. AND CARTER, M. W. (1971). "A rapid liquid scintillation technique for monitoring ^{14}C in air," page 381 in *Rapid Methods for Measuring Radioactivity in Environment* (International Atomic Energy Agency, Vienna).

MOORE, F. (1960). *Liquid-Liquid Extraction with High Molecular Weight Amines,* Report NAS-NS-3101 (National Technical Information Service, U. S. Department of Commerce, Springfield, Virginia).

MOORE, F. L. (1968). "New extraction chromatographic method for rapid separation of americium from other transuranium elements," Anal. Chem. **40**, 2130.

MORGAN, A. AND MITCHELL, G. R. (1962). *A New Method for the Determination of Iodine-131 in Herbage Including Measurements Made in Autumn, 1961,* Report AERE-M-1004 (U. K. Atomic Energy Research Establishment, Harwell, England).

MORIUCHI, S. AND MIYANAGA, I. (1966). "A spectrometric method for measurement of low-level gamma exposure dose," Health Physics **12**, 541.

MORKEN, D. A. (1959). "Integral-phosphor counting of alpha emitters," Health Physics **2**, 77.

MORRISON, G. H. AND FRIESER, H. (1957). *Solvent Extraction in Analytical Chemistry* (John Wiley and Sons, New York).

MOSES, H., STEHNEY, A. F. AND LUCAS, H. F., JR. (1960). "The effects of meteorological variables upon the vertical and temporal distributions of atmospheric radon," J. Geophys. Res. **65,** 1233.

MURTHY, G. K. AND CAMPBELL, J. E. (1959). "A method for the rapid ashing of milk for radionuclide analysis," J. of Dairy Sci. **42,** 1288.

MURTHY, G. K. AND CAMPBELL, J. E. (1966). "In vitro transformation of inorganic iodine to protein-bound iodine in milk by formaldehyde," J. of Dairy Sci. **49,** 1190.

MYERS, W. A. AND LINDNER, M. (1971). "Precise determination of the natural abundance of ^{237}Np and ^{239}Pu in Katanga pitchblende," J. Inorg. Nucl. Chem. **33,** 3233.

NAKASHIMA, Y. AND KAWANO, M. (1972). "Gamma-ray dosimetry by means of proportional counter, (I) Fundamental characteristics of plastic wall counter," J. Nucl. Sci. Tech. **9,** 47 [Atomic Energy Society of Japan].

NAS-NRC (1959). National Academy of Sciences – National Research Council, *Radioactive Waste Disposal from Nuclear-Powered Ships,* Publication 658 (National Academy of Sciences – National Research Council, Washington).

NAS-NRC (1960–71). National Academy of Sciences – National Research Council, *Radiochemistry of the Elements,* Reports NAS-NS-3001 through -3058: *Radiochemical Techniques*, Reports NAS-NS-3101 through -3113 (National Technical Information Service, U. S. Department of Commerce, Springfield, Virginia).

NAS-NRC (1962). National Academy of Sciences – National Research Council, *Disposal of Low-Level Radioactive Waste into the Pacific Coastal Waters,* Publication 985 (National Academy of Sciences – National Research Council, Washington).

NAS-NRC (1964). National Academy of Sciences – National Research Council, *Studies in the Penetration of Charged Particles in Matter,* Nuclear Science Series Report 39 (National Academy of Sciences – National Research Council, Washington).

NAS-NRC (1971). National Academy of Sciences – National Research Council, *Radioactivity in the Marine Environment* (National Academy of Sciences – National Research Council, Washington).

NAS-NRC (1972). National Academy of Sciences – National Research Council, *The Effects on Populations of Exposure to Low Levels of Ionizing Radiation* (National Academy of Sciences – National Research Council, Washington).

NAS-NRC (1973). National Academy of Sciences – National Research Council, *Radionuclides in Food* (National Academy of Sciences – National Research Council, Washington).

NCRP (1954). National Council on Radiation Protection and Measurements, *Permissible Dose from External Sources of Ionizing Radiation,* NCRP Report No. 17, published as National Bureau of Standards Handbook 59 (U. S. Government Printing Office, Washington).

NCRP (1959). National Council on Radiation Protection and Measurements, *Maximum Permissible Body Burdens and Maximum Permissible Concentrations of Radionuclides in Air and in Water for Occupational Exposure,*

NCRP Report No. 22, published as National Bureau of Standards Handbook 69 (U. S. Government Printing Office, Washington).

NCRP (1961). National Council on Radiation Protection and Measurements, *Stopping Powers for Use with Cavity Chambers,* NCRP Report No. 27, published as National Bureau of Standards Handbook 79 (U. S. Government Printing Office, Washington).

NCRP (1971). National Council on Radiation Protection and Measurements, *Basic Radiation Protection Criteria,* NCRP Report No. 39 (National Council on Radiation Protection and Measurements, Washington).

NCRP (1975). National Council on Radiation Protection and Measurements, *Natural Background Radiation in the United States,* NCRP Report No. 45 (National Council on Radiation Protection and Measurements, Washington).

NCRP (1976). National Council on Radiation Protection and Measurements, *Tritium Measurement Techniques,* NCRP Report No. 47 (National Council on Radiation Protection and Measurements, Washington).

NCRP (1977). National Council on Radiation Protection and Measurements, forthcoming NCRP report entitled *A Manual of Radioactivity Measurements Procedures* (National Council on Radiation Protection and Measurements, Washington).

NEGRO, V. C., WATNICK, S. AND RAFT, P. D. (1974). "A temperature-compensated electrometer for environmental measurements," IEEE Trans. on Nucl. Sci. **NS-21,** 805.

NEHER, H. V. (1958). "The primary cosmic radiation," Ann. Rev. Nucl. Sci. 8, 217.

NEHER, H. V. (1967). "Cosmic-ray particles that changed from 1954 to 1958 to 1965," J. Geophys. Res. **72,** 1527.

NEHER, H. V. (1971). "Cosmic rays at high latitudes and altitudes covering four solar maxima," J. Geophys. Res. **76,** 1637.

NEHER, H. V. AND ANDERSON, H. R. (1965). "Cosmic ray changes during a solar cycle," page 153 in *Proc. of Ninth Intl. Conf. on Cosmic Rays,* Vol. 1 (Institute of Physics and the Physical Society, London).

NEILER, J. H. AND BELL, P. R. (1965). "The scintillation method," α, β, γ Spectroscopy **1,** 284.

NELEPO, B. A. (1960). "Gamma-spectrometric measurements of the radioactivity of the waters of the Atlantic Ocean," Herald of Moscow University 5, 36 [Translation No. 6838 from Joint Publications Research Service, U. S. Department of Commerce, Washington].

NELSON, D. J., Ed. (1971). *Radionuclides in Ecosystems,* Report CONF-710501-PI (Oak Ridge National Laboratory, Oak Ridge, Tennessee).

NELSON, D. J. AND EVANS, F. C., Eds. (1969). *Symposium on Radioecology,* Report CONF-670503 (U. S. Atomic Energy Commission, Washington).

NERURKAR, N. AND WEBBER, W. R. (1964). "Observations of primary cosmic-ray variations using ion chamber and geiger counters," J. Geophys. Res. **69,** 815.

NEWSTEIN, H., COHEN, L. D. AND KRABLIN, R. (1971). "An automated atmospheric radon sampling system," Atmos. Environ. **4,** 823.

NIELSEN, J. M. (1972). "Gamma ray spectrometry," page 678 in *Physical Methods of Chemistry,* Vol. 1, Part III D, Weissberger, A. and Rossiter, B. W., Eds. (John Wiley and Sons, New York).

NIELSEN, J. M. AND BEASLEY, T. M. (1967). "Radiochemical determination of plutonium for radiological purposes," page 923 in *Plutonium Handbook,* Vol. 2, Wick, O. J., Ed. (Gordon and Breach, New York).

NIELSEN, J. M. AND STAFF MEMBERS (1973). *Pacific Northwest Laboratory Annual Report for 1972; Vol. 2, Physical Sciences, Part 2, Radiological Sciences,* Report BNWL-1751, Part 2 (Battelle Northwest Laboratories, Richland, Washington).

NOAKES, J. E., NEARY, M. P. AND SPAULDING, J. D. (1973). "Tritium measurements with a new liquid scintillation counter," Nucl. Instrum. Meth. **109,** 177.

NOVIKOV, YU. P., MYASOEDOV, B. F. AND IVANOVA, S. A. (1972). "Neptunium extraction from alkaline solutions," Radiochem. Radioanal. Letters **9,** 85.

NOZAKI, Y. AND TSUNAGAI, S. (1973). "A simultaneous determination of lead-210 and polonium-210 in sea water," Anal. Chim. Acta **64,** 209.

NUCLEAR DATA GROUP (1965). *Nuclear Data Sheets, 1959–1965, Part 5, A = 119 through A = 137* (Academic Press, New York).

OAKLEY, D. T. (1972). *Natural Radiation Exposure in the United States,* Report ORP/SID 72-1 (U. S. Environmental Protection Agency, Washington).

O'BRIEN, K. (1972). "The cosmic ray field at ground level," page 15 in *The Natural Radiation Environment II,* Adams, J. A. S., Lowder, W. M. and Gesell, T., Eds., Report CONF-720805 (U. S. Energy Research and Development Administration, Washington).

O'BRIEN, K. AND MCLAUGHLIN, J. E. (1972). "The radiation dose to man from galactic cosmic rays," Health Physics **22,** 225.

O'BRIEN, K., LOWDER, W. M. AND SOLON, L. R. (1958). "Beta and gamma dose rates from terrestrially distributed sources," Radiat. Res. **9,** 216.

O'DELL, A. A. (1971). *Monte Carlo Analysis of Air-to-Ground Correlation Parameters for Finite Distributed Gamma Sources,* Report S-546-R, EGG-1183-2258 (E. G. & G. Inc., Santa Barbara, California).

O'DELL, A. A. (1972). "Monte Carlo simulation of aerial radiation measurements for distributed ground-based gamma sources," page 619 in *The Natural Radiation Environment II,* Adams, J. A. S., Lowder, W. M. and Gesell, T., Eds., Report CONF-720805 (U. S. Energy Research and Development Administration, Washington).

OESCHGER, H. (1963). "Low-level counting methods," page 13 in *Radioactive Dating* (International Atomic Energy Agency, Vienna).

O'KELLEY, G. D. (1968). *Detection and Measurement of Nuclear Radiation,* Report NAS-NS 3105 (National Academy of Sciences—National Research Council, Washington).

OLSON, D. (1973). Private communication (U. S. Atomic Energy Commission, Health Services Laboratory, Idaho Falls, Idaho).

OSBORNE, R. V. AND HILL, C. R. (1964). "High resolution alpha particle spectroscopy at very low specific activities," Nucl. Instrum. Meth. **29,** 101.

OSTLUND, H. G., MASON, A. S. AND YDFALK, A. (1972). *Tritium Laboratory Data Report #2, Atmospheric HT and HTO, 1968–1971*, Data Report No. 2 (University of Miami, Miami, Florida).

OVERMAN, R. J. AND CLARK, H. M. T. (1960). *Radioisotope Techniques* (McGraw-Hill, New York).

OWERS, M. J. (1959). *The Concentration by Electrodialysis of Caesium and Strontium Radioactivities in Waters*, Report AERE R-3010 (U. K. Atomic Energy Research Establishment, Harwell, England).

PANCHOLI, S. C. AND MARTIN, M. J. (1972). "Nuclear data sheets for A = 212," Nucl. Data Sheets **B8**, 165.

PARKER, H. M. (1946). *Review of Air Monitoring Procedures at Clinton Laboratories*, Report MDDC-471 (U. S. Atomic Energy Commission, Oak Ridge, Tennessee).

PASTERNACK, B. S. AND HARLEY, J. H. (1971). "Detection limits for radionuclides in the analysis of multi-component gamma-ray spectrometer data," Nucl. Instrum. Meth. **91**, 533.

PATHAK, P. N. AND SARABHAI, V. (1970). "A study of the long-term modulation of galactic cosmic ray intensity," Planetary and Space Sci. **18**, 81.

PATTERSON, R. L., JR. AND LOCKHART, L. B., JR. (1964). "Geographical distribution of lead-210 (RaD) in the ground-level air," page 383 in *The Natural Radiation Environment*, Adams, J. A. S. and Lowder, W. M., Eds. (University of Chicago Press, Chicago).

PAULY, J., GUZZI, G., GIRARDI, F. AND BORELLA, A. (1966). "Application of gamma ray spectrometry and computer techniques to the determination of the minimum detectable content of trace elements in neutron activated materials," Nucl. Instrum. Meth. **42**, 15.

PECK, E. L. (1972). "Snow measurement predicament," Water Resources Res. **8**, 244.

PECK, E. L. AND BISSELL, V. C. (1971). "Aerial measurement of snow water equivalent by terrestrial gamma radiation survey," presented at the XV Int. Union of Geodesy Geophysics (General Assembly, Moscow, USSR).

PECK, E. L., BISSELL, V. C., JONES, E. B. AND BURGE, D. L. (1971). "Evaluation of snow water equivalent by airborne measurement of passive terrestrial gamma radiation," Water Resources Res. **7**, 1151.

PEIRSON, D. H. (1963). "Beryllium-7 in air and rain," J. Geophys. Res. **68**, 3831.

PEIRSON, D. H., CAMBRAY, R. S. AND SPICER, G. S. (1966). "Lead-210 and polonium-210 in the atmosphere," Tellus **18**, 427.

PELLETIER, C. A. (1970). "Performance and design of an environmental survey," page 55 in *Environmental Surveillance in the Vicinity of Nuclear Facilities*, Reinig, W. C., Ed. (C. C. Thomas, Springfield, Illinois).

PENSKO, J. (1967). "Variations of the gamma-ray background dose rates and atmospheric radon concentration in Warsaw," Atompraxis **13**, 310.

PENSKO, J., GWIAZDOWSKI, B., JAGICLAK, J., BIERNACKA, M. AND MAMONT, K. (1971). "Combined environmental radioactivity measurements for rapid estimation of the gamma-radiation field," page 443 in *Rapid Methods for*

Measuring Radioactivity in the Environment (International Atomic Energy Agency, Vienna).

PEPPARD, D. F. (1971). "Liquid-liquid extraction of the actinides," Ann. Rev. Nucl. Sci. **21**, 365.

PEPPARD, D. F., MASON, G. W. AND ANDREJASICH, C. M. (1963). "Applicability of certain organo-phosphorus extractants to environmental and biological radioactivity surveys," J. Inorg. Nucl. Chem. **25**, 1175.

PERKINS, R. W. (1961). "A large well crystal for the direct measurement of trace amounts of radioisotopes in environmental samples," Health Physics **7**, 81.

PERKINS, R. W. (1963). "Physical and chemical form of I^{131} in fallout," Health Physics **9**, 1113.

PERKINS, R. W. (1965). "An anticoincidence shielded multidimensional gamma-ray spectrometer," Nucl. Instrum. Meth. **33**, 71.

PERKINS, R. W. (1969). "A procedure for the continuous separation and subsequent direct counting of short-lived cosmic-ray produced radionuclides in rainwater," page 79 in *Pacific Northwest Laboratory Report for 1968,* Vol. 2, Pt. 2, Nielsen, J. M., Ed., Report BNWL-1051 (Battelle Northwest Laboratories, Richland, Washington).

PERKINS, R. W. AND RANTICELLI, L. A. (1971). *Nuclear Techniques for Trace Elements and Radionuclide Measurements in Natural Waters,* Report BNWL-SA-3993 (Battelle Northwest Laboratories, Richland, Washington).

PERKINS, R. W., NIELSEN, J. M. AND DIEBEL, R. N. (1960). "Total absorption gamma-ray spectrometers utilizing anticoincidence shielding," Rev. Sci. Instrum. **31**, 1344.

PERTSOV, L. A. (1973). *Ioniziruyushchie Izlucheniya Biosfery* [Ionizing Radiation in the Biosphere] (Atomizdat, Moscow).

PETROW, H. G. (1965). "Rapid determination of strontium-90 in bone ash via solvent extraction of yttrium-90," Anal. Chem. **37**, 584.

PETROW, H. G. AND COVER, A. (1965). "Direct radiochemical determinations of lead-210 in bone," Anal. Chem. **37**, 1659.

PETROW, H. G., COVER, A., SCHIESSLE, W. AND PARSONS, E. (1964). "Radiochemical determination of radium-228 and thorium-228 in biological and mineral samples," Anal. Chem. **36**, 1600.

PHELPS, P. L., HAMBY, K. O., SHORE, B. AND POTTER, G. D. (1968). *Ge(Li) Gamma-Ray Spectrometers of High Sensitivity and Resolution for Biological and Environmental Counting,* Report UCRL-50437 (Lawrence Livermore Laboratory, Livermore, California).

PHILBIN, T. W. AND WHIPPLE, G. H. (1970). "An environmental survey for the Nine Mile Point Power Reactor," page 93 in *Environmental Surveillance in the Vicinity of Nuclear Facilities,* Reinig, W. C., Ed. (C. C Thomas, Springfield, Illinois).

PHS (1967). U. S. Public Health Service, *Radioassay Procedures for Environmental Samples* (National Center for Radiological Health, Rockville, Maryland).

PIESCH, D. (1968). "Routine dosimetry with phosphate glasses," page 830 in

Luminescence Dosimetry, Auxier, J. A., Becker, K. and Robinson, E. M., Eds., Report CONF-680920 (U. S. Atomic Energy Commission, Washington).

PITKIN, J. A. (1968). "Airborne measurements of terrestrial radioactivity as an aid to geologic mapping," U. S. Geologic Professional Paper 516-F (U. S. Geological Survey, Washington).

PLATO, P. A. AND GELSKY, D. (1972). "The use of field instruments in the operation of environmental surveys around nuclear power plants," Water Quality Instrum. **1**, 176.

POLIKARPOV, G. G. (1966). *Radioecology of Aquatic Organisms: the Accumulation and Biological Effect of Radioactive Substances* (Reinhold Publishing Corp., New York).

PORTER, C. R., AUGUSTINE, R. J., MATUSZEK, J. M., JR. AND CARTER, M. W. (1965). *Procedures for Determination of Stable Elements and Radionuclides in Environmental Samples,* Public Health Service Publication No. 999-RH-10 (U. S. Department of Health, Education and Welfare, Washington).

PRICE, W. J. (1958). *Nuclear Radiation Detection* (McGraw-Hill Book Co., New York).

PROCTOR, C. M., LOUREY, A. R., PAPADOPULOS, E. AND PENNING, J. R., JR. (1962). "Response of gamma scintillation detectors for field survey use," Limnology and Oceanography **7**, 273.

PUPHAL, K. W. AND OLSEN, D. R. (1972). "Electrodeposition of alpha-emitting nuclides from a mixed oxalate-chloride electrolyte," Anal. Chem. **44**, 284.

PURVIS, A. E. AND BUCKMEIER, F. J. (1969). "Comparison of airborne spectral gamma radiation data with field verification of measurement," page 553 in *Proc. Sixth Intl. Symp. on Remote Sensing of Environment* (University of Michigan, Ann Arbor, Michigan).

PUSHKAREV, V. V., SKRYLEV, L. D. AND BAGRETSOV, F. (1960). "Concentration of radioactive cesium using gelatine foam formation," Radiochemistry **1**, 219 [Translation of Radiochimiya (USSR)].

PUTNAM, M. H., GIPSON, D. H., HELMER, R. G. AND HEATH, R. L. (1965). *A Nonlinear Least-Square Program for the Determination of Parameters of Photopeaks by the Use of a Modified-Gaussian Function,* Report IDO-17016 (U. S. Atomic Energy Commission, Idaho Falls, Idaho).

RAFT, P. D., LOWDER, W. M. AND BECK, H. L. (1970). *Measurements of Cosmic Ray Ionization in the Atmosphere, 1968–70,* Report HASL-234 (U. S. Atomic Energy Commission, New York).

RAO, U. R. (1972). "Solar modulation of galactic cosmic radiation," Space Sci. Rev. **12**, 719.

REYNOLDS, S. A. (1962). "Methods and standards in the analysis of water," Health Physics **8**, 391.

REINIG, W. C., Ed. (1970). *Environmental Surveillance in the Vicinity of Nuclear Facilities* (C. C. Thomas, Springfield, Illinois).

RIECK, H. G. (1967). "Air sampling equipment at remote locations," page 26 in *Pacific Northwest Laboratory Annual Report for 1966,* Vol. 2, Part 2,

Nielsen, J. M., Ed., Report BNWL-481 (Battelle Northwest Laboratories, Richland, Washington).

RIEL, G. K. (1966). *Concentration of Radioactive Isotopes in Environmental Water Measured by Underwater Gamma Spectrometry,* Report NOLTR 65-23 (Naval Surface Weapons Center, Silver Spring, Maryland).

RIEL, G. K. AND SIMONS, D. G. (1962). "Underwater cosmic ray measurement," J. Geophys. Res. **67,** 891.

RIEL, G. K., MADIGOSKY, S. W. AND SIMONS, D. G. (1969). *Gamma Spectrum Data Processing Programs,* Report NOLTR 69-233 (Naval Surface Weapons Center, Silver Spring, Maryland).

RIEL, G. K., HAWKINS, W., LIEBLER, G. AND DUFFEY, D. (1973). "Radioactivity monitor for power plant environmental water," Nucl. Tech. **17,** 275.

ROBERTSON, D. E. (1973). "Determination of silver in seawater by anion exchange and instrumental neutron activation analysis," page 24 in *Pacific Northwest Laboratory Annual Report for 1972,* Vol. 2, Part 2, Nielsen, J. M., Ed., Report BNWL-1751 (Battelle Northwest Laboratories, Richland, Washington).

ROBINSON, C. V. (1967). "Geiger-Müller and proportional counters," page 57 in *Instrumentation in Nuclear Medicine,* Vol. 1, Hine, G. J., Ed. (Academic Press, New York).

RODRIGUEZ-PASQUÉS, R. H., STEINBERG, H. L., HARDING, J. E., MULLER, P. A., HUTCHINSON, J. M. R. AND MANN, W. R. (1972). "Low-level radioactivity measurements on aluminum, steel and copper," Int. J. Appl. Rad. and Isotopes **23,** 445.

ROLLE, R. (1970). "Determination of radon daughters on filters by simple liquid scintillation technique," Am. Indust. Hyg. Assoc. J. **31,** 718.

ROSE, J. E. AND SHONKA, F. R. (1968). "Calculated calibrations for ion chambers fabricated from plastics simulating air and muscle; determination of W and Γ_{Ra}," Radiat. Res. **36,** 384.

ROSHOLT, J. W. (1957). "Quantitative radiochemical methods for determination of the sources of natural radioactivity," Anal. Chem. **29,** 1398.

ROTH, S. J. AND HUCKABAY, G. (1974). "A comparison of the in situ measurement of terrestrial Americium with a Ge(Li) spectrometer and a FIDLER," IEEE Trans. on Nucl. Sci. **NS-21,** 438.

RYTZ, A. (1973). "Catalogue of recommended alpha energy and intensity values," Atomic Data and Nuclear Data Tables **12,** 479.

SAKANOUE, M. AND TSUJI, T. (1971). "Plutonium content of soil at Nagasaki," Nature **234,** 92.

SANDSTROM, A. E. (1965). *Cosmic Ray Physics* (North Holland Publishing Co., Amsterdam).

SANSONI, B. AND KRACKE, W. (1971). "Rapid determination of low-level biological materials using wet ashing by ·OH radicals," page 217 in *Rapid Methods for Measuring Radioactivity in the Environment* (International Atomic Energy Agency, Vienna).

SAX, N. I., DENNY, J. D. AND REEVES, B. R. (1968). "Modified scintillation counting technique for determination of low level krypton-85," Anal. Chem. **40,** 1915.

SCHELL, W. R. (1971). "Gas counting of tritium," page 113 in *Tritium*, Moghissi, A. A. and Carter, M. W., Eds. (Messenger Graphics, Phoenix, Arizona).

SCHMIDT, C. T. AND KOCH, J. J. (1966). "Plutonium survey with x-ray detectors," page 1 in *Hazards Control Progress Report No. 26*, Report UCRL-50007-66-2 (Lawrence Livermore Laboratory, Livermore, California).

SCHMORAK, M. R. (1970a). "Nuclear data sheets for A = 232," Nucl. Data Sheets **B4**, 561.

SCHMORAK, M. R. (1970b). "Nuclear data sheets for A = 240," Nucl. Data Sheets **B4**, 661.

SCHULMAN, J. H. (1967). "Survey of luminescence dosimetry," page 3 in *Luminescence Dosimetry*, Symposium Series 8, Attix, F. H., Ed. (U. S. Atomic Energy Commission, Washington).

SCHULMAN, J. H., GINTHER, R. J., KLICK, C. C., ALGER, R. S. AND LENG, R. A. (1951). "Dosimetry of x-rays and gamma-rays by radiophotoluminescence," J. Appl. Sci. **22**, 1479.

SCHULTZ, V. AND KLEMENT, A. W., JR., Eds. (1963). *Radioecology* (Reinhold Publishing Corp., New York).

SCHUMACHER, E. (1957). "Über fokussierenden Ionenaustausch IV. Zur Theorie des Fokussierungseffektes," Helv. Chim. Acta **40**, 2322.

SCHWARZER, T. F., COOK, B. G. AND ADAMS, J. A. S. (1972). "Low altitude gamma-spectrometric surveys from helicopters in Puerto Rico as an example of the remote sensing of thorium, uranium and potassium in rocks and soils," Remote Sensing of Environment **2**, 83.

SCHWENDIMAN, L. C. AND HEALY, J. W. (1958). "Nuclear-track technique for low-level Pu in urine," Nucleonics **16** (6) 78.

SEDLET, J. (1966). "Radon and radium," page 219 in *Treatise on Analytical Chemistry*, Vol. 4, Part 2, Kolthoff, I. M. and Elving, P. J., Eds. (John Wiley and Sons, New York).

SEDLET, J. AND STEHNEY, A. F. (1974). Personal Communication (Argonne National Laboratory, Argonne, Illinois).

SETTER, L. R., ANDREW, R., CLEMAN, R., FRIEND, A., MARKARIAN, C. AND STORG, A. (1966). *Routine Surveillance of Radioactivity Around Nuclear Facilities*, Report PHS-799-RH-23 (U. S. Public Health Service, Washington).

SEYMOUR, A. H. (1971). "Introduction," page 1 in *Radioactivity in the Marine Environment* (National Academy of Sciences—National Research Council, Washington).

SHAMBON, A. (1972). *Some Implications of a Laboratory Study of LiF Dosimeters for Environmental Radiation Measurements*, Report HASL-251 (U. S. Atomic Energy Commission, New York).

SHAMOS, M. H. AND LIBOFF, A. R. (1966). "A new measurement of the intensity of cosmic-ray ionization at sea level," J. Geophys. Res. **71**, 4651.

SHAMOS, M. H. AND LIBOFF, A. R. (1968). "New ionization chamber technique for the measurement of environmental radiation," Rev. Sci. Instrum. **39**, 223.

SHAMOS, M. H., LIBOFF, A. R. AND SIDEROWITZ, J. (1964). "Thin wall ion chambers for low-level radiation environment," page 627 in *The Natural Radiation Environment,* Adams, J. A. S. and Lowder, W. M., Eds. (University of Chicago Press, Chicago).

SHANNON, L. V. AND ORREN, M. J. (1970). "A rapid method for the determination of polonium-210 and lead-210 in sea water," Anal. Chim. Acta **52,** 166.

SHAPIRO, J., SEGALL, A. AND WORCESTER, J. (1968). "The application of a personnel monitoring system to population dosimetry of background radiation," page 625 in *Proc. of First Intl. Conf. on Radiation Protection* (Pergamon Press, New York).

SHEA, M. F. AND SMART, D. F. (1967). "Worldwide trajectory-derived vertical cutoff rigidities and their application to experimental measurements," J. Geophys. Res. **72,** 2021.

SHEA, M. F., SMART, D. F., MCCRACKEN, K. G. AND RAO, U. R. (1968). *Suppl. IQSY Instruction Manual 10 Cosmic Ray Tables,* Special Report 71, AFCRL 68-0030 (Air Force Cambridge Research laboratories, Bedford, Massachusetts).

SHEEHY, J. P. (1965). *Chemical Analysis of Environmental Radionuclides* (Robert A. Taft Sanitary Engineering Center, U. S. Department of Health, Education and Welfare, Cincinnati, Ohio).

SHERWOOD, J. D., MOOR, D. T. AND KURODA, P. K. (1973). "Radioactive cerium isotopes: the fallout from recent French and Chinese nuclear weapons tests," Health Physics **24,** 491.

SHONKA, F. R. (1962). "Vibrating quartz fiber electrometer," Radiology **78,** 112.

SHOPAUSKAS, K. K. (1964). "Vertical distribution of radon daughter products in the troposphere as a function of meteorological conditions," Publ. Sci. Acad. Lithuanian SSR, Series B, **1,** 21.

SHUPING, R. E., PHILLIPS, C. R. AND MOGHISSI, A. A. (1969). "Low level counting of environmental krypton-85 in liquid scintillation," Anal. Chem. **41,** 2082.

SILKER, W. B. (1972). "The physical form of radionuclides in Pacific Ocean water," page 20 in *Pacific Northwest Laboratory Annual Report for 1971,* Vol. 2, Part 2, Nielsen, J. M., Ed., Report BNWL-1651 (Battelle Northwest Laboratories, Richland, Washington).

SILKER, W. B., PERKINS, R. W. AND RIECK, H. G. (1971). "A sampler for concentrating radionuclides from natural water," Ocean Engineering **2,** 49.

SILL, C. W. (1961). "Decomposition of refractory silicates in ultramicro analysis," Anal. Chem. **33,** 1684.

SILL, C. W. (1969). "Separation and radiochemical determination of uranium and the transuranium elements using barium sulfate," Health Physics **17,** 89.

SILL, C. W. AND WILLIS, C. P. (1964). "Precipitation of submicrogram quantities of thorium by barium sulfate and application to fluorometric determination of thorium in mineralogical and biological samples," Anal. Chem. **36,** 622.

SILL, C. W. AND WILLIS, C. P. (1965). "Radiochemical determination of ^{210}Pb in mill products and biological materials," Anal. Chem. **37,** 1661.

SILL, C. W. AND WILLIAMS, R. L. (1969). "Radiochemical determination of uranium and the transuranium elements in process solutions and environmental samples," Anal. Chem. **41,** 1624.

SILL, C. W. AND WILLIAMS, R. L. (1971). "Rapid identification and determination of alpha emitters in environmental samples," page 201 in *Rapid Methods for Measuring Radioactivity in the Environment* (International Atomic Energy Agency, Vienna).

SILL, C. W., PUPHAL, K. W. AND HINDMAN, F. D. (1974). "Simultaneous determination of alpha-emitting nuclides of radium through californium in soil," Anal. Chem. **46,** 1725.

SIMPSON, J. A. AND WANG, J. R. (1970). "The eleven-year and residual solar modulation of cosmic rays (1952–1969)," Astrophys. J. **161,** 265.

SMITH, D. G., COCHRAN, J. A. AND SHLEIEN, B. (1970). *Calibration and Initial Field Testing of ^{85}Kr Detectors for Environmental Monitoring,* Report BRH/NERHL 70-4 (U. S. Public Health Service, Rockville, Maryland).

SMITH, K. A. AND MERCER, E. R. (1970). "The determination of radium-226 and radium-228 in soils and plants using radium-225 as a yield tracer," J. Radioanal. Chem. **5,** 303.

SNYDER, W. F., FORD, M. R., WARNER, G. G. AND FISHER, H. L., JR. (1969). "Estimates of absorbed fractions for monoenergetic photon sources uniformly distributed in various organs of a heterogeneous phantom," J. Nucl. Medicine, Suppl. No. 3, Pamphlet 5.

SOLON, L. R., LOWDER, W. M., ZILA, A. V., BLATZ, H. AND EISENBUD, M. (1959). "External environmental radiation measurements in the United States," page 159 in *Proc. of Second Intl. Conf. on Peaceful Uses of Atomic Energy* (Pergamon Press, New York).

SOMAYAJULU, B. L. K., LAL, D. AND CRAIG, H. (1973). "Silicon-32 profiles in the South Pacific," Earth and Planetary Sci. Letters **18,** 181.

SPENCER, L. V. AND ATTIX, F. H. (1955). "A theory of cavity ionization," Radiat. Res. **3,** 239.

SPIERS, F. W. AND OVERTON, T. R. (1962). "Attenuation factors for certain tissues when the body is exposed to nearly omni-directional gamma radiation," Phys. in Med. and Biol. **7,** 35.

SPIERS, F. W., MCHUGH, M. J. AND APPLEBY, D. B. (1964). "Environmental γ-ray dose to populations, surveys made with a portable meter," page 885 in *The Natural Radiation Environment,* Adams, J. A. S. and Lowder, W. M., Eds. (University of Chicago Press, Chicago).

SPITZ, H. B. AND WRENN, M. E. (1974). "The diurnal variation of the radon-222 concentration in residential structures in Grand Junction, Colorado," page 100 in *Second Workshop on the Natural Radiation Environment,* Lowder, W. M., Ed., Report HASL-287 (U. S. Energy Research and Development Administration, New York).

STALEY, D. O. (1966). "The diurnal oscillations of radon and thoron and their decay products," J. Geophys. Res. **71,** 3357.

STEIN, L. (1973a). "Removal of radon from air by oxidation with bromine trifluoride," J. Inorg. Nucl. Chem. **35,** 1973.

STEIN, L. (1973b). "Removal of xenon and radon from contaminated atmospheres with dioxygenyl hexafluoroantimonate, O_2SbF_6," Nature 243, 30.

STEPHENS, L. D. AND CONTELOW, H. (1975). *Annual Environmental Monitoring Report—1974,* Report LBL-3821 (U. S. Energy Research and Development Administration, Washington).

STEVENSON, D. L. AND JOHNS, F. B. (1971). "Separation technique for the determination of ^{85}Kr in the environment," page 157 in *Rapid Methods for Measuring Radioactivity in the Environment* (International Atomic Energy Agency, Vienna).

STORRUSTE, A., REISTAD, A., RUDJORD, T., DAHLER, A. AND LIESTOL, I. (1965). "Measurement of environmental gamma radiation on Norwegian houses," Health Physics **11,** 261.

STUART, T. P. (1971). "Use of aerial surveys for determining plutonium contamination," page 328 in *Proc. of the Rocky Flats Symp. on Safety in Plutonium Handling Facilities,* Report CONF-710401 (U. S. Atomic Energy Commission, Washington).

STUART, T. P., HENDRICKS, T. J., GUDIKSEN, P. H., JONES, D. E. AND LYNCH, O. D. T. (1973). "An aerial radiological survey of Eniwetok Atoll," Trans. Amer. Nucl. Soc. **16,** 540.

STUDIER, M. H., POSTMUS, C., JR., MECH, J., WALTERS, R. R. AND SLOTH, E. N. (1962). "The use of ^{129}I as an isotopic tracer and its determination along with normal ^{127}I by neutron activation—the isolation of iodine from a variety of materials," J. Inorg. Nucl. Chem. **24,** 755.

SUGIHARA, T. T. (1961). *Low-Level Radiochemical Separations,* Report No. NAS–NS 3103 (National Academy of Sciences—National Research Council, Washington).

SUNDERMAN, D. N. AND MEINKE, W. W. (1957). "Evaluation of radiochemical separation procedures," Anal. Chem. **29,** 1578.

SUTTON, D. C. AND KELLY, J. J. (1968). "^{90}Sr and ^{137}Cs measurements of large volume sea water samples," Report HASL-196 (U. S. Atomic Energy Commission, New York).

TALVITIE, N. A. (1972). "Electrodeposition of actinides for alpha spectrometric determination," Anal. Chem. **44,** 280.

Technical Manager (Chemistry) (1961). *Analytical Methods for the Determination of Iodine-131 in Vegetation, Milk, Thyroid Glands, and Natural Waters,* PG Report 204(W) (U. K. Atomic Energy Authority, Risley, England).

TERADA, K., HAYAKAWA, H., SAWADA, K. AND KIBA, T. (1970). "Silica gel as a support for inorganic ion-exchangers for the determination of cesium-137 in natural waters," Talanta **17,** 955.

THOMAS, C. W. (1973). "Analytical procedure for measuring ^{218}Po, ^{214}Pb, and ^{214}Bi in rain samples using Ge(Li) diodes," page 32 in *Pacific Northwest Laboratory Annual Report for 1972,* Vol. 2, Part 2, Nielsen, J. M., Ed., Report BNWL-1751 (Battelle-Northwest Laboratories, Richland, Washington).

THOMAS, C. W., YOUNG, J. A., PERKINS, R. W. AND WOGMAN, N. A. (1969). "Analytical procedures used for measuring cosmic-ray induced radionuclides ^{7}Be, ^{24}Na, ^{34m}Cl, ^{38}Cl, and ^{39}Cl in rain and air," page 76 in *Pacific Northwest Laboratory Annual Report for 1968,* Vol. 2, Part 2, Nielsen, J. M., Ed., Report BNWL-1051 (Battelle Northwest Laboratories, Richland, Washington).

THOMAS, C. W., PERKINS, R. W. AND YOUNG, J. A. (1971). "Analytical procedure for cosmogenic radionuclides in rain water," page 31 in *Pacific Northwest Laboratory Annual Report for 1970,* Vol. 2, Part 2, Nielsen, J. M., Ed., Report BNWL-1551 (Battelle-Northwest Laboratories, Richland, Washington).

THOMAS, J. W. (1968). *Determination of Radon Progeny in Air from Alpha Activity in Air Samples,* Report HASL-202 (U. S. Atomic Energy Commission, New York).

THOMAS, J. W. (1972). "Measurement of radon daughters in air," Health Physics **23,** 783.

THOMPSON, T. AND WIBERG, P. A. (1963). "Some observations of variations of the natural background radiation," Tellus **15,** 313.

TINNEY, J. F., KOCH, J. J. AND SCHMIDT, C. T. (1969). *Plutonium Survey with an X-Ray Sensitive Detector,* Report UCRL-71362 (Lawrence Livermore Laboratory, Livermore, California).

UNSCEAR (1966). United Nations Scientific Committee on the Effects of Atomic Radiation, *Report of the United Nations Scientific Committee on the Effects of Atomic Radiation* (United Nations, New York).

UNSCEAR (1972). United Nations Scientific Committee on the Effects of Atomic Radiation, *Ionizing Radiation: Levels and Effects,* A Report of the United Nations Scientific Committee on the Effects of Atomic Radiation to the General Assembly with Annexes (United Nations, New York).

USAEC (1973a). United States Atomic Energy Commission, *Final Environmental Statement Concerning Proposed Rule Making Action: Numerical Guides for Design Objectives and Limiting Conditions for Operation to Meet the Criterion "As Low as Practicable," for Radioactive Material in Light-Water-Cooled Nuclear Power Reactor Effluents,* Report WASH-1258 (U. S. Atomic Energy Commission, Washington).

USAEC (1973b). United States Atomic Energy Commission, "Measurements of radionuclides in the environment—analysis of I-131 in milk," Regulatory Guide 4.3 (U. S. Atomic Energy Commission, Washington).

USNRC (1975). United States Nuclear Regulatory Commission, Title 10, Part 50, Appendix I, Code of Federal Regulations, Federal Register **40,** 19439.

U. S. STANDARD ATMOSPHERE (1962). Report AD-401813 (U. S. Government Printing Office, Washington).

VDOVENKO, V. M. (1967). *Coprecipitation and Adsorption of Radioactive Elements,* Report AEC-tr-6716 (NTIS, U. S. Department of Commerce, Springfield, Virginia).

VERHEUL, H. (1971). "Nuclear data sheets for A = 87," Nucl. Data Sheets **B5,** 457.

VILENSKII, V. D., DAVYDOV, E. N. AND MALAKHOV, S. G. (1965). "On the seasonal and geographic variations in lead-210 content in the atmosphere," page 89 in *Radioactive Isotopes in the Atmosphere and Their Use in Meteorology,* I. L. Karol' *et al.,* Eds. (Atomizdat, Moscow; English Translation: Israel Program for Scientific Translations, Jerusalem, 1967).

VINOGRADOV, A. P. (1959). *The Geochemistry of Rare and Dispersed Chemical Elements in Soils,* 2nd ed. (Consultants Bureau, New York).

VINOGRADOV, A. P., Ed. (1966 to present). *Analytical Chemistry of the Elements* (Vernadskii Institute on Geochemistry and Analytical Chemistry, USSR Academy of Sciences) [English Translation by Israel Program for Scientific Translations, Jerusalem, from Ann Arbor-Humphrey Science Publishers, Ann Arbor, Michigan].

VOILLEQUE, P. G. AND BALDWIN, B. R., Eds. (1971). *Health Physics Aspects of Nuclear Facility Siting* (Health Physics Society, Idaho Falls, Idaho).

VOLCHOK, H. L., FEINER, M., SIMPSON, H. J., BROECKER, W. S., NOSHKIN, V. E., BOWEN, V. T. AND WILLIS, E. (1970). "Ocean fallout-the Crater Lake experiment," J. Geophys. Res. **75,** 1084.

WADA, M. (1960). "Atmospheric effects on the intensity of cosmic-ray mesons." Sci. Papers Inst. Phys. Chem. Res. (Japan) **54,** 335.

WAHL, A. C. AND BONNER, N. A. (1951). *Radioactivity Applied to Chemistry* (John Wiley and Sons, New York).

WATT, D. E. AND RAMSDEN, D. (1963). *High Sensitivity Counting Techniques* (Pergamon Press, Oxford, and MacMillan Co., New York).

WATARI, K., IMAI, K. AND IZAWA, M. (1971). "Multi-loaded metal salt(s) – ion exchange resins: their preparation and application," J. of Japan Health Soc. **6,** 7.

WEAST, R. C., Ed. (1971). *Handbook of Chemistry and Physics,* 52nd ed. (Chemical Rubber Publishing Co., Cleveland, Ohio).

WELLER, R. I. (1964). "Low-level radioactive contamination," page 567 in *The Natural Radiation Environment,* Adams, J. A. S. and Lowder, W. M., Eds. (University of Chicago Press, Chicago).

WEBBER, W. R. (1967). "The spectrum and charge composition of the primary cosmic radiation," page 181 in *Encyclopedia of Physics,* Vol. 46, No. 2, Sitte, K., Ed. (Springer-Verlag, New York).

WENSEL, A. (1964). "Measurement of environmental radiation with plastic scintillators," page 643 in *The Natural Radiation Environment,* Adams, J. A. S. and Lowder, W. M., Eds. (University of Chicago Press, Chicago).

WESSMAN, R. A., MAJOR, W. J., LEE, K. D. AND LEVENTHAL, L. (1971). "Commonality in water, soil, air, vegetation, and biological sample analysis for plutonium," page 73 in *Proceedings of Environmental Plutonium Symposium,* Fowler, E. B., Henderson, R. W. and Milligan, M. F., Eds., Report LA-4756 (Los Alamos Scientific Laboratory, Los Alamos, New Mexico).

WHITE, F. A. AND COLLINS, T. L. (1954). "A two stage magnetic analyzer for isotopic ratio determination of 10^4 to 1 or greater," Appl. Spectroscopy **8,** 169.

WHITE, F. A. AND FORMAN, L. (1967). "Four-stage mass spectrometer of large radius," Rev. Sci. Instrum. **38,** 355.

WHITE, F. A., ROURKE, F. M. AND SHEFFIELD, J. C. (1958). "A three stage mass spectrometer," Appl. Spectroscopy **12,** 46.

WHO (1966). World Health Organization, *Methods of Radiochemical Analysis* (World Health Organization, Geneva).

WHO (1968). World Health Organization, *Routine Surveillance for Radionuclides in Air and Water* (World Health Organization, Geneva).

WILLIAMS, D., CAMBRAY, R. S. AND MASKELL, S. C. (1959). *An Airborne Radiometric Survey of the Windscale Area, October 19–22, 1957,* Report AERE-R2890 (Atomic Energy Research Establishment, Harwell, England).

WILSON, C. R. AND CAMERON, J. R. (1968). "Dosimetric properties of $Li_2B_4O_7$:Mn (Harshaw)," page 161 in *Proc. of Second Intl. Conf. on Luminescence Dosimetry,* Report CONF-680920 (U. S. Atomic Energy Commission, Washington).

WILSON, E. B., JR. (1952). *An Introduction of Scientific Research* (McGraw-Hill, New York).

WINTER, M. (1971). "Moglichkeiten zur Messung der akkumulierten Strahlungsdosis in der Umgebung kerntechnischer Anlagen," page 525 in *Rapid Methods for Measuring Radioactivity in the Environment* (International Atomic Energy Agency, Vienna).

WOGMAN, N. A. (1975). Personal Communication (Battelle Northwest Laboratories, Richland, Washington).

WOLLENBERG, H. A. AND SMITH, A. R. (1964). "Studies in terrestrial γ radiation," page 513 in *The Natural Radiation Environment,* Adams, J. A. S. and Lowder, W. M., Eds. (University of Chicago Press, Chicago).

WRENN, M. E., JINKS, S. M., COHEN, N. AND HAIRR, G. M. (1974). "Cesium-137 and cesium-134 distributions in sediments, water, and biota of the lower Hudson River and their dosimetric implications for man," page 279 in *Proc. of Third Intl. Congress of the International Radiation Protection Association,* Report CONF-730907-P1 (U. S. Atomic Energy Commission, Washington).

YAMASHITA, T., NADA, N., ONISHI, H. AND KITAMURA, S. (1971). "Calcium sulfate activated thulium or dysprosium for thermoluminescence dosimetry," Health Physics **21,** 295.

YANIV, S. S., BRODSKY, A. AND SASHIN, D. (1972). "Analysis of Pu-Am spectra in a Xe proportional counter: effect of characteristic x-rays," Health Physics **23,** 705.

YASYULENIS, R. YU., LUYANAS, V. YU. AND KIKITE, V. P. (1973). "Method of determining sodium-24 in atmospheric precipitation," Soviet Radiochemistry **14,** 673.

YEATES, D. B., GOLDIN, A. S. AND MOELLER, D. W. (1972). "Natural radiation in the urban environment," Nuclear Safety **13,** 275.

ZIMMERMAN, D. W., RHYNER, C. R. AND CAMERON, J. R. (1966). "Thermal annealing effects on the thermoluminescence of LiF," Health Physics **12,** 525.

ZOTIMOV, N. V. (1968). "Investigations of a method of measuring snow storage by using the gamma radiation of the earth," State Hydrological Institute (Trudy GGI) **152,** 114.

The NCRP

The National Council on Radiation Protection and Measurements is a nonprofit corporation chartered by Congress in 1964 to:

1. Collect, analyze, develop, and disseminate in the public interest information and recommendations about (a) protection against radiation and (b) radiation measurements, quantities, and units, particularly those concerned with radiation protection;
2. Provide a means by which organizations concerned with the scientific and related aspects of radiation protection and of radiation quantities, units, and measurements may cooperate for effective utilization of their combined resources, and to stimulate the work of such organizations;
3. Develop basic concepts about radiation quantities, units, and measurements, about the application of these concepts, and about radiation protection;
4. Cooperate with the International Commission on Radiological Protection, the International Commission on Radiation Units and Measurements, and other national and international organizations, governmental and private, concerned with radiation quantities, units, and measurements and with radiation protection.

The Council is the successor to the unincorporated association of scientists known as the National Committee on Radiation Protection and Measurements and was formed to carry on the work begun by the Committee.

The Council is made up of the members and the participants who serve on the fifty-four Scientific Committees of the Council. The Scientific Committees, composed of experts having detailed knowledge and competence in the particular area of the Committee's interest, draft proposed recommendations. These are then submitted to the full membership of the Council for careful review and approval before being published.

The following comprise the current officers and membership of the Council:

Officers

President	LAURISTON S. TAYLOR
President-Elect	WARREN K. SINCLAIR
Vice President	E. DALE TROUT

Secretary and Treasurer	W. Roger Ney
Assistant Secretary	Eugene R. Fidell
Assistant Treasurer	Harold O. Wyckoff

Members

Seymour Abrahamson
Roy E. Albert
John A. Auxier
William J. Bair
Victor P. Bond
Harold S. Boyne
Robert L. Brent
A. Bertrand Brill
Reynold F. Brown
William W. Burr, Jr.
Melvin W. Carter
George W. Casarett
Randall S. Caswell
Arthur B. Chilton
Stephen Cleary
Cyril L. Comar
Patricia Durbin
Merril Eisenbud
Thomas S. Ely
Benjamin G. Ferris
Asher J. Finkel
Donald C. Fleckenstein
Richard F. Foster
Hymer L. Friedell
Arthur H. Gladstein
Robert A. Goepp
Marvin Goldman
Robert O. Gorson
Arthur W. Guy
Ellis M. Hall
John H. Harley
Robert J. Hasterlik
John W. Healy
John M. Heslep
Marylou Ingram
Seymour Jablon
Jacob Kastner
Edward B. Lewis
Charles W. Mays
Roger O. McClellan
Mortimer Mendelsohn
Dade W. Moeller
Russell H. Morgan
Paul E. Morrow
Robert D. Moseley, Jr.
James V. Neel
Robert J. Nelsen
Peter C. Nowell
Frank Parker
Herbert M. Parker
Chester R. Richmond
Lester Rogers
Harald H. Rossi
Robert E. Rowland
William L. Russell
John H. Rust
Eugene L. Saenger
Harry F. Schulte
Raymond Seltser
Warren K. Sinclair
Walter S. Snyder
Lewis V. Spencer
J. Newell Stannard
Chauncey Starr
John B. Storer
Lauriston S. Taylor
Roy C. Thompson
E. Dale Trout
Arthur C. Upton
John C. Villforth
George L. Voelz
Niel Wald
Edward W. Webster
George M. Wilkening
Harold O. Wyckoff

Honorary Members

Edgar C. Barnes
Carl B. Braestrup
Austin M. Brues
Frederick P. Cowan
Robley D. Evans
Paul C. Hodges
George V. LeRoy
Karl Z. Morgan
Edith H. Quimby
Shields Warren

Currently, the following Scientific Committees are actively engaged in formulating recommendations:

SC-1: Basic Radiation Protection Criteria
SC-7: Monitoring Methods and Instruments
SC-11: Incineration of Radioactive Waste
SC-18: Standards and Measurements of Radioactivity for Radiological Use
SC-22: Radiation Shielding for Particle Accelerators
SC-23: Radiation Hazards Resulting from the Release of Radionuclides into the Environment
SC-24: Radionuclides and Labeled Organic Compounds Incorporated in Genetic Material
SC-25: Radiation Protection in the Use of Small Neutron Generators
SC-26: High Energy X-Ray Dosimetry
SC-28: Radiation Exposure from Consumer Products
SC-30: Physical and Biological Properties of Radionuclides
SC-31: Selected Occupational Exposure Problems Arising from Internal Emitters
SC-32: Administered Radioactivity
SC-33: Dose Calculations
SC-34: Maximum Permissible Concentrations for Occupational and Non-Occupational Exposures
SC-37: Procedures for the Management of Contaminated Persons
SC-38: Waste Disposal
SC-39: Microwaves
SC-40: Biological Aspects of Radiation Protection Criteria
SC-41: Radiation Resulting from Nuclear Power Generation
SC-42: Industrial Applications of X Rays and Sealed Sources
SC-44: Radiation Associated with Medical Examinations
SC-45: Radiation Received by Radiation Employees
SC-46: Operational Radiation Safety
SC-47: Instrumentation for the Determination of Dose Equivalent
SC-48: Apportionment of Radiation Exposure
SC-50: Surface Contamination
SC-51: Radiation Protection in Pediatric Radiology and Nuclear Medicine Applied to Children
SC-52: Conceptual Basis of Calculations of Dose Distributions
SC-53: Biological Effects and Exposure Criteria for Radiofrequency Electromagnetic Radiation
SC-54: Bioassay for Assessment of Control of Intake of Radionuclides

In recognition of its responsibility to facilitate and stimulate cooperation among organizations concerned with the scientific and related aspects of radiation protection and measurement, the Council has created a category of NCRP Collaborating Organizations. Organizations or groups of organizations which are national or international in scope and are concerned with scientific problems involving radiation quantities, units, measurements and effects, or radiation protection may be admitted to collaborating status by the Council. The present Collaborating Organizations with which the NCRP maintains liaison are as follows:

American Academy of Dermatology
American Association of Physicists in Medicine

American College of Radiology
American Dental Association
American Industrial Hygiene Association
American Insurance Association
American Medical Association
American Nuclear Society
American Occupational Medical Association
American Podiatry Association
American Public Health Association
American Radium Society
American Roentgen Ray Society
American Society of Radiologic Technologists
American Veterinary Medical Association
Association of University Radiologists
Atomic Industrial Forum
College of American Pathologists
Defense Civil Preparedness Agency
Genetics Society of America
Health Physics Society
National Bureau of Standards
National Electrical Manufacturers Association
Radiation Research Society
Radiological Society of North America
Society of Nuclear Medicine
United States Air Force
United States Army
United States Energy Research and Development Administration
United States Environmental Protection Agency
United States Navy
United States Nuclear Regulatory Commission
United States Public Health Service

The NCRP has found its relationships with these organizations to be extremely valuable to continued progress in its program.

The Council's activities are made possible by the voluntary contribution of the time and effort of its members and participants and the generous support of the following organizations:

Alfred P. Sloan Foundation
American Academy of Dental Radiology
American Academy of Dermatology
American Association of Physicists in Medicine
American College of Radiology
American College of Radiology Foundation
American Dental Association
American Industrial Hygiene Association
American Insurance Association
American Medical Association
American Mutual Insurance Alliance
American Nuclear Society
American Occupational Medical Association
American Osteopathic College of Radiology
American Podiatry Association
American Public Health Association
American Radium Society
American Roentgen Ray Society
American Society of Radiologic Technologists
American Veterinary Medical Association
American Veterinary Radiology Society
Association of University Radiologists
Atomic Industrial Forum
Battelle Memorial Institute
College of American Pathologists
Defense Civil Preparedness Agency
Edward Mallinckrodt, Jr. Foundation
Genetics Society of America
Health Physics Society
James Picker Foundation
National Association of Photographic Manufacturers
National Bureau of Standards
National Electrical Manufacturers Association
Radiation Research Society
Radiological Society of North America
Society of Nuclear Medicine
United States Energy Research and Development Administration
United States Environmental Protection Agency
United States Public Health Service

To all of these organizations the Council expresses its profound appreciation for their support.

Initial funds for publication of NCRP reports were provided by a grant from the James Picker Foundation and for this the Council wishes to express its deep appreciation.

The NCRP seeks to promulgate information and recommendations based on leading scientific judgment on matters of radiation protection and measurement and to foster cooperation among organizations concerned with these matters. These efforts are intended to serve the public interest and the Council welcomes comments and suggestions on its reports or activities from those interested in its work.

NCRP Reports

NCRP Reports are distributed by the NCRP Publications' office. Information on prices and how to order may be obtained by directing an inquiry to:

NCRP Publications
P.O. Box 30175
Washington, D.C. 20014

The extant NCRP Reports are listed below.

NCRP Report No.	Title
8	*Control and Removal of Radioactive Contamination in Laboratories* (1951)
9	*Recommendations for Waste Disposal of Phosphorus-32 and Iodine-131 for Medical Users* (1951)
10	*Radiological Monitoring Methods and Instruments* (1952)
12	*Recommendations for the Disposal of Carbon-14 Wastes* (1953)
14	*Protection Against Betatron-Synchrotron Radiations Up To 100 Million Electron Volts* (1954)
16	*Radioactive Waste Disposal in the Ocean* (1954)
22	*Maximum Permissible Body Burdens and Maximum Permissible Concentrations of Radionuclides in Air and in Water for Occupational Exposure* (1959) [Includes Addendum 1 issued in August 1963]
23	*Measurement of Neutron Flux and Spectra for Physical and Biological Applications* (1960).
25	*Measurement of Absorbed Dose of Neutrons and of Mixtures of Neutrons and Gamma Rays* (1961)
27	*Stopping Powers for Use with Cavity Chambers* (1961)
28	*A Manual of Radioactivity Procedures* (1961)
30	*Safe Handling of Radioactive Materials* (1964)
31	*Shielding for High-Energy Electron Accelerator Installations* (1964)
32	*Radiation Protection in Educational Institutions* (1966)
33	*Medical X-Ray and Gamma-Ray Protection for Energies Up to 10 MeV—Equipment Design and Use* (1968)
35	*Dental X-Ray Protection* (1970)

36	*Radiation Protection in Veterinary Medicine* (1970)
37	*Precautions in the Management of Patients Who Have Received Therapeutic Amounts of Radionuclides* (1970)
38	*Protection Against Neutron Radiation* (1971)
39	*Basic Radiation Protection Criteria* (1971)
40	*Protection Against Radiation From Brachytherapy Sources* (1972)
41	*Specification of Gamma-Ray Brachytherapy Sources* (1974)
42	*Radiological Factors Affecting Decision-Making in a Nuclear Attack* (1974)
43	*Review of the Current State of Radiation Protection Philosophy* (1975)
44	*Krypton-85 in the Atmosphere—Accumulation, Biological Significance, and Control Technology* (1975)
45	*Natural Background Radiation in the United States* (1975)
46	*Alpha-Emitting Particles in Lungs* (1975)
47	*Tritium Measurement Techniques* (1976)
48	*Radiation Protection for Medical and Allied Health Personnel* (1976)
49	*Structural Shielding Design and Evaluation for Medical Use of X Rays and Gamma Rays of Energies Up to 10 MeV* (1976)
50	*Environmental Radiation Measurements* (1976)

Binders for NCRP Reports are available. Two sizes make it possible to collect into small binders the "old series" of reports (NCRP Reports Nos. 8–31) and into large binders the more recent publications (NCRP Reports Nos. 32–50). Each binder will accommodate from five to seven reports. The binders carry the identification "NCRP Reports" and come with label holders which permit the user to attach labels showing the reports contained in each binder.

The following bound sets of NCRP Reports are also available:

Volume I. NCRP Reports Nos. 8, 9, 10, 12, 14, 16, 22
Volume II. NCRP Reports Nos. 23, 25, 27, 28, 30, 31
Volume III. NCRP Reports Nos. 32, 33, 35, 36, 37
Volume IV. NCRP Reports Nos. 38, 39, 40, 41
Volume V. NCRP Reports Nos. 42, 43, 44, 45, 46

(Titles of the individual reports contained in each volume are given above.)

The following NCRP reports are now superseded and/or out of print:

NCRP Report No.	Title
1	*X-Ray Protection* (1931). [Superseded by NCRP Report No. 3]

2 *Radium Protection* (1934). [Superseded by NCRP Report No. 4]
3 *X-Ray Protection* (1936). [Superseded by NCRP Report No. 6]
4 *Radium Protection* (1938). [Superseded by NCRP Report No. 13]
5 *Safe Handling of Radioactive Luminous Compounds* (1941). [Out of print]
6 *Medical X-Ray Protection up to Two Million Volts* (1949). [Superseded by NCRP Report No. 18]
7 *Safe Handling of Radioactive Isotopes* (1949). [Superseded by NCRP Report No. 30]
11 *Maximum Permissible Amounts of Radioisotopes in the Human Body and Maximum Permissible Concentrations in Air and Water* (1953). [Superseded by NCRP Report No. 22]
13 *Protection Against Radiations from Radium, Cobalt-60 and Cesium-137* (1954). [Superseded by NCRP Report No. 24]
15 *Safe Handling of Cadavers Containing Radioactive Isotopes* (1953). [Superseded by NCRP Report No. 21]
17 *Permissible Dose from External Sources of Ionizing Radiation* (1954) including *Maximum Permissible Exposure to Man, Addendum to National Bureau Of Standards Handbook 59* (1958). [Superseded by NCRP Report No. 39]
18 *X-Ray Protection* (1955). [Superseded by NCRP Report No. 26]
19 *Regulation of Radiation Exposure by Legislative Means* (1955). [Out of print]
20 *Protection Against Neutron Radiation Up to 80 Million Electron Volts* (1957). [Superseded by NCRP Report No. 38]
21 *Safe Handling of Bodies Containing Radioactive Isotopes* (1958) [Superseded by NCRP Report No. 37]
24 *Protection Against Radiations from Sealed Gamma Sources* (1960). [Superseded by NCRP Reports Nos. 33, 34 and 40]
26 *Medical X-Ray Protection Up to Three Million Volts* (1961). [Superseded by NCRP Reports Nos. 33, 34, 35 and 36]
29 *Exposure to Radiation in an Emergency* (1962). [Superseded by NCRP Report No. 42]
34 *Medical X-Ray and Gamma-Ray Protection for Energies Up to 10 MeV—Structural Shielding Design and Evaluation* (1970). [Superseded by NCRP Report No. 49].

The following statements of the NCRP were published outside of the NCRP Report series:

"Blood Counts, Statements of the National Committee on Radiation Protection," Radiology 63, 428 (1954)

"Statements on Maximum Permissible Dose from Television Receivers and Maximum Permissible Dose to the

Skin of the Whole Body," Am. Jr. of Roentgenol., Radium Therapy and Nucl. Med. 84, 152 (1960) and Radiology 75, 122 (1960)

X-Ray Protection Standards for Home Television Receivers, Interim Statement of the National Council on Radiation Protection and Measurements (National Council on Radiation Protection and Measurements, Washington, 1968)

Specification of Units of Natural Uranium and Natural Thorium (National Council on Radiation Protection and Measurements, Washington, 1973)

Copies of the statements published in journals may be consulted in libraries. A limited number of copies of the last two statements listed above are available for distribution by NCRP Publications.

Index*

* Boldface entries refer to principal discussion of a topic.